T0210650

Synthesis Lectures on Mechanical Engineering

This series publishes short books in mechanical engineering (ME), the engineering branch that combines engineering, physics and mathematics principles with materials science to design, analyze, manufacture, and maintain mechanical systems. It involves the production and usage of heat and mechanical power for the design, production and operation of machines and tools. This series publishes within all areas of ME and follows the ASME technical division categories.

Issam Abu-Mahfouz

Instrumentation: Theory and Practice, Part 1

Principles of Measurements

 Springer

Issam Abu-Mahfouz
Pennsylvania State University Harrisburg
Middletown, PA, USA

ISSN 2573-3168   ISSN 2573-3176 (electronic)
Synthesis Lectures on Mechanical Engineering
ISBN 978-3-031-15248-1   ISBN 978-3-031-15246-7 (eBook)
https://doi.org/10.1007/978-3-031-15246-7

This Springer imprint is published by the registered company Springer Nature Switzerland AG
The registered company address is: Gewerbestrasse 11, 6330 Cham, Switzerland

Preface

In today's world, measurements and sensors are fundamental to all facets of our daily life. This Part I of 'Instrumentation: Theory and Practice' emphasizes simple and concise coverage of the fundamental aspects of measuring systems. It is designed to provide the reader with essential knowledge regarding signals, signal analysis, signal conditioning circuits, and data acquisition systems.

This textbook is intended for use as an introductory one semester course at the junior level of an undergraduate program. It is also very relevant for technicians, engineers, and researchers who had no formal training in instrumentation and wish to engage in experimental measurements. The prerequisites are a basic knowledge of multivariable calculus, introductory physics, and a familiarity with basic electrical circuits and components.

The book chapters are organized as follows; Chap. 1 introduces the measuring process and the basics of standards in the field of measurements. A short section on smart sensors is also presented. Chapter 2 discusses the common types of signals and the dynamics of measuring systems' responses. It concludes with the Fourier transform and frequency spectrum analysis of time signals. An important task in measurements is the analysis of the collected data and the quantification of the measurement uncertainties. These topics are covered in Chap. 3 with emphasis on types of errors and the use of statistical analysis. Topics of signal connectivity, signal conditioning, and processing using analog circuits such as the Wheatstone bridge and op-amps are presented in Chap. 4. Chapter 5 is dedicated to discussing fundamentals of transistors, digital logic circuits with applications for both combinational and sequential logic. Data acquisition systems (DASs) and the digitization of analog signals are presented in Chap. 6 with emphasis on analog-to-digital converters (ADCs).

Each chapter includes illustrative figures, relevant electric circuits, and is complemented with well-designed examples and problems. I hope the reader will find this book clear and useful as a first step in understanding the principles of measurements.

Middletown, PA, USA   Issam Abu-Mahfouz

Acknowledgments I would very much like to acknowledge and thank my family for their encouragement, patience, and support. I would also like to thank and recognize the administration at Penn State Harrisburg for the encouragement and for granting me the sabbatical time to help me complete this book. I am particularly grateful to Executive Editor Mr. Paul Petralia and the entire staff at Springer Nature for their consistent encouragement and support.

I would appreciate and welcome feedback comments about the book by writing directly to the author at iaa2@psu.edu.

Contents

About the Author

Issam Abu-Mahfouz, Ph.D., P.E. is an Associate Professor of Mechanical Engineering and Chair of the Mechanical Engineering Program, School of Science, Engineering, and Technology, at Penn State University Harrisburg, where he has been working since 1999. He received his Bachelor of Science and Master's degrees in Mechanical Engineering from Kuwait University in 1985 and 1989, respectively, and his Doctorate degree in Mechanical Engineering from Case Western Reserve University in 1993. He has taught courses in computer-aided design (CAD), finite element analysis (FEA), capstone design, automatic controls, mechatronics, instrumentation, fluid power, design for manufacturability, dynamics, vibrations, optimization, and smart systems. Before joining Penn State, he served in the industry as a project engineer (design and manufacture of heavy trenching machines) and as an R&D engineer (metal drilling). He provides engineering consulting and research services for the industry in the design and development of new products. His research interests include manufacturing process and machine condition monitoring, machinery noise and vibration, and the application of artificial intelligence techniques.

Part I
Principles of Measurements

Introduction

1

1.1   The Measuring Process

Measurement is the process of associating a number with a physical variable or a quantity under measurement (QUM). Taking a measurement involves interaction between the target object and the measuring instrument. Based on the measuring objectives, the variable to be measured, and the conditions at the target, this interaction may require direct contact between the object and sensor, or it may be conducted remotely (noncontact) using different techniques [1, 2]. This interaction may also involve some sort of energy exchange between the object and the sensor instrument. It is important to ensure that the output of the measuring instrument is an accurate and reliable representation of the true value within acceptable tolerance.

Measurements are integrated in almost every aspect of our daily life. They can be conducted manually or are automatically performed continuously or intermittently for different purposes. These include measurements for basic research, product test and development, health care, food industry, security, and defense, etc. Fig. 1.1 shows a generalized flow diagram of the measuring process.

Internal and external noise and interferences can affect the measuring system performance. This may result in corrupted or distorted signals and reduce the value of the collected data. Depending on the QUM external errors include electromagnetic interference, radiation, and background noise or dynamic noise associated with the object or process being measured. Whereas internal errors are inherent to the measuring instrumentation and may include linearity, precision, hysteresis, and accuracy. These errors are discussed in Chapter 2.

Measuring systems also include interface circuits or modules, for serial or parallel, digital input–output data communication. Wireless signal transmission over local or remote networks is becoming very common. Analysis may need to be performed on acquired

© The Author(s), under exclusive license to Springer Nature Switzerland AG 2022
I. Abu-Mahfouz, *Instrumentation: Theory and Practice, Part 1*, Synthesis Lectures
on Mechanical Engineering, https://doi.org/10.1007/978-3-031-15246-7_1

Fig. 1.1 A flow diagram for a general measuring system

signals for feature extraction and pattern recognition. In addition, the information should be presented in a form that can be easily interpreted to facilitate reaching a conclusion or a decision for the intended measurement project. Measurement systems also include means for storage and display of collected data and results of analysis.

Measuring instruments are designed and fabricated using several components each with its own intended function. In the following, three examples are discussed briefly to illustrate the measuring mechanism. More details of these and many other measuring devices are covered in Part II of this book [3].

1.1.1 Tire Pressure Gauge

Figure 1.2 illustrates the main components of a tire pressure gauge. A small piston can slide inside the gauge tube which has a smooth polished surface. To minimize friction and reduce resistance to motion, the inside tube surface is slightly lubricated. When pressure is applied at one side of the piston the spring is compressed until its force balanced the tire air force on the piston. At this point the piston is in static equilibrium and the following relations apply

$$\text{spring force} = \text{air force}$$

$$k_{spring}.\Delta x_{piston} = p_{tire}.A_{piston}$$

$$\Delta x_{indicator} = \frac{p_{tire}.A_{piston}}{k_{spring}}$$

$$\text{Transfer Function} = \text{Sensitivity}$$

$$= \frac{\Delta x_{\max}}{p_{tire,full\,range}} = \frac{A_{piston}}{k_{spring}} \tag{1.1}$$

where,

$A_{piston} = $ piston circular area.

Fig. 1.2 Pencil tire pressure
gauge

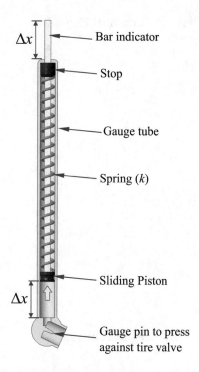

k_{spring} = spring stiffness constant (N/m) or (lb/inch).

$\Delta x_{spring} = \Delta x_{indicator}$ = piston displacement due pressure.

p_{tire} = tire air pressure being measured.

A calibrated bar or rod indicator with marked scale allows reading of the pressure.
The calibrated bar fits inside the gauge spring (with no contact) and is not attached to the
piston. However, it is pushed outside the piston by the piston when pressure is applied.
When the pressure is released, the spring will push the piston back to home (0 pressure)
position, however, the rod keep its maximum position to allow for the reading to be
recorded. As can be clearly deduce from Eq. (1.1) that the sensitivity of the tire gauge
can be determined from the selection of the piston diameter and the spring stiffness (k).
For a certain gauge tube (fixed A_{piston}) a softer spring (smaller k) will compress more and
cause further extension out of the indicator bar (piston displacement Δx) per unit gauge
pressure.

Fig. 1.3 Potentiometer types
and circuit

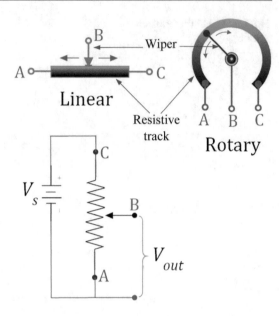

1.1.2 Potentiometers

A potentiometer is a variable resistor circuit that produces a voltage output V_{out} that is proportional to the change in resistance. Figure 1.3 illustrates the potentiometer circuit and the linear and rotation types of this device. The resistive track is a conductive material that is formed in a straight strip or a in a circular arc shape. The track has a uniform resistivity along its length, so it can give a linear variation of resistance proportional to the motion of the wiper. The potentiometer acts like a linear variable voltage divider. This simple device can be used for linear and angular position or displacement measurements.

1.1.3 Liquid-in-Glass Thermometer

This is probably the most used temperature measuring devices. Figure 1.4 shows the major features of the liquid-in-glass thermometer. The bulb holds a specified amount of the measuring liquid and is fully in contact with the object or fully immersed in the fluid for which temperature measurement is sought. with the object or which is usually alcohol or mercury. Alcohol and mercury are the most sued measuring liquids for this device. When the liquid in the bulb and in the capillary tube exchange heat with the target through the glass body (bulb and stem), it expands or contracts upon heating or cooling respectively. Therefore, the level of the liquid column will rise or fall in the capillary tube. The stem is marked (scaled) with the temperature readings though calibration. Keeping all other design parameters, the same, a higher coefficient of thermal expansion leads to

Fig. 1.4 Liquid-in-glass
thermometer

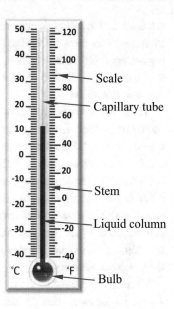

more expansion or contraction for the same change in temperature which is give better
readability such as for alcohol. However, mercury has a higher boiling temperature than
alcohol and is used to measure much higher temperature.

1.2   Standards

Among other standards in science and engineering, measurements standards cover units,
testing methods, calibration, and materials. A standard is a known accurate physical representation of a quantity (unit) of measurement and is agreed upon throughout the world.
Standards are established to facilitate uniformity in conducting experiments and exchang-
ing data worldwide with minimal confusion and ambiguity. The use of standards ensures
high quality, reliable and reproducible measurements.

A dimension is a fundamental measurable physical variable used to quantify or
describe a physical variable. Examples are length, temperature, mass, time, electric current, light intensity, and substance. A unit is the quantitative measure (magnitude) of
its associate dimension. Examples are meter, degrees Kelvin, kilogram, second, ampere,
candela, and mole. These are the seven base units defined by the international system
of units (SI) and all other units can be derived from these base units [4, 5]. A unit is
realized by reference or comparison to a material standard or to a physical phenomenon,
including physical and atomic constants. For example, the fundamental unit of length in
the international system (SI) is the meter, defined as the distance travelled by light in a
vacuum during a time interval of 1/299,792,458 of a second [5, 6]. The inch-pound units

are used in the US (US customary units) and they are related to the SI system of units by conversion factors. Modern standard units of measurements are defined in terms of fundamental physical constants. This eliminate the need for inter-comparison of artifacts which are susceptible to wear and deterioration limiting the precision and stability of the standard definition.

There are several institutes and organizations that are responsible for maintaining standards at their localities across the world. The General Conference on Weights and Measures (CGPM) is the international agency responsible for maintaining exact uniform standards of measurements. **NIST** (The National Institute for Standards and Technology) is responsible for developing, maintaining, and disseminating national standards in USA. NIST is also responsible for assessing the measurement uncertainties associated with the values assigned to their standards. NIST also provides standard reference materials (SRM) that are used to perform instrument calibrations and measurements accuracy verification. In addition, NIST participates in promoting uniformity in international legal metrology laws and enforcement.

There are different levels of standards with varying degree of accuracy. The following are the major categories of measurements standards.

Primary Standards: These are the ultimate reference standards of the units of measurements they represent and are made to the highest metrological quality. They are defined by international agreements and are maintained at the national standards laboratories worldwide. They are fixed, preserved under careful conditions, and are protected by law. Primary standards are not available for use outside these laboratories. The main purpose of these reference standards is to calibrate and verify the secondary standards.

Secondary Standards: These are the closest to by Secondary or calibration standards are regularly compared and certified against the primary standards by their related industrial laboratories. Secondary standards are mainly used to calibrate and verify the tertiary standards by national and certified calibration laboratories.

Tertiary Standards: These are reference standards that are used to inspect and calibrate the working standards for workshops and laboratories.

Working Standards: These standards are used to periodically calibrate workshop and laboratory instruments and measuring equipment to ensure proper accuracy and performance. Working standards must be traceable to the secondary and primary standards. For example, gauge blocks are used as a working standard to calibrate measurement tools such as calipers and micrometers.

Calibration: Establishes a relation between quantity values (i.e. measurement value) against reference or standard values by comparison under controlled conditions and following standard procedures. It also results in specifying associated measurement uncertainties. Calibration establishes the relation (characteristic or sensitivity) between the input (physical variable) and the output (indication) of a measuring device. Accuracy of calibration is limited to the accuracy of the standard used. NIST provides calibration services to disseminate the primary physical measurement standards for the U.S. commerce, industry, and research.

In addition to the fundamental standards on units, standards are also available for other engineering and manufacturing measurement and instrumentation products. Most of these standards are created and maintained through agreements or consensus by standard committees established by professional organizations and societies such as the American Society of Testing and Materials (ASTM), the American Society of Mechanical Engineers (ASME), and the American National Standards Institute (ANSI). These standards are, in general, voluntary and they provide best practices in the respective fields. However, some standards such as test codes, are enforced or regulated by government agencies. The following is a sample list of standards produced by the above mentioned organizations.

- ASME Boiler and Pressure Vessel Code
- ASME Y14.5 Dimensioning an Tolerancing
- Data Acquisition Systems (PTC 19.22)
- Temperature Measurement (PTC 19.3)
- Measurement of Fluid Flow in Pipes Using Vortex Flowmeters (MFC-6)
- Measurement of Shaft Power (PTC 19.7)
- Electrical Power Measurements (PTC 19.6)
- Pressure Measurement (PTC 19.2)
- Measurement Uncertainty and Conformance Testing: Risk Analysis (An ASME Technical Report) (B89.7.4.1)
- Measurement of Fluid Flow in Pipes Using Orifice, Nozzle, and Venturi (MFC-3 M)
- Measurement of Industrial Noise (PTC 36)
- IEC 60,751 Ed. 3.0 b:2022—Industrial Platinum Resistance Thermometers and Platinum Temperature Sensors
- 1451.1–1999—Standard for a smart transducer interface for sensors and actuators— Network capable application processor (NCAP) information model.

In general, standards facilitate interoperability between technologies and components made by different manufacturers.

1.3   Smart Sensors

Smart sensors are measurement devices that are integrated with onboard microprocessors, micro peripherals, and standard communication interfaces. These are required to enable smart sensors to accomplish noise filtering, signal amplification, data conversion, signals processing, data analysis, and data transfer. In addition to collecting measurements of physical variables, they can be programmed to automatically process signals and perform predefined control and data transfer tasks. For example, in home automation, smart sensors can aid in monitoring and controlling light, climate (HVAC), temperature, appliances, door and window locks, water temperature, and other building mechanical and electrical systems. This will increase comfort and reduce energy consumption.

Smart sensors are usually connected to networks via communication protocols and are vital elements in enabling the 'Internet of Things' (IoT). The IoT encompasses almost any device or entity that can be instrumented with an integrated circuit device, given a unique identity (IP), and capable of transmitting signals over a network of connected devices or over the internet. Web enabled and connected devices may include sensors, actuators, communication hubs, servers, computers (stationary or mobile). Smart Sensors can be designed to perform self-diagnosis and self-calibration. Some smart sensors are designed with multi-sensing capabilities of different variables within one package.

1.4   Examples

Example 1.1 To verify the opration of a flow rate transducer, the experiminter measured the time period to fill a 1.5 gallons container to be equal to 50.7 s. Based on this data, determine the flow rate of water. Convert the answer to cubic meters per hour and also calculate the mass flow rate in kilograms per minute.

Solution:

Volume flow rate is given by

$$Q = \frac{\Delta V}{\Delta t} = \frac{1.5}{50.7} = 0.0296 \, \text{gal/s}$$

Using conversion factors the flow rate in m^3/h is

$$Q = \left(\frac{0.0296}{1}\right)\left(\frac{3.7854 \times 10^{-3} \, m^3}{1}\right)\left(\frac{3600}{1\,h}\right) = 0.4034 \, m^3/h$$

Take density of water as 1000 kg/m, we get

$$Q = \left(\frac{0.4034}{1}\right)\left(\frac{1000 \, \text{kg}}{1}\right)\left(\frac{1}{60 \, \text{min}}\right) = 6.723 \, \text{kg/min}$$

Few digits are kept for intermediate calculations to reduce the accumulation of truncation or round-off error in the final result. Final result could be reported appropriately as 6.7 kg/min with only two significant digits (limited by $\Delta V = 1.5$).

Example 1.2 How many significant digits are in each of the following numbers? Write each in scientific notation.

Solution:

Given data	Number of significant digits	Scientific notation
53.8	3	5.38×10^1
16.030	5	1.6030×10^1
49,600	5	4.9600×10^4
3000	1	3×10^3
0.0052	2	5.2×10^{-3}
0.100	3	1.00×10^{-1}

Example 1.3 Round each of the following number to the designated significant digits.

Solution:

Given data	Round to this number of significant digits	Rounded number
53.8	2	54
16.030	3	16.0
2.625134	3	2.63
5000	1	5×10^3
0.00575	2	5.8×10^{-3}
3.850	2	3.8

1.5   Problems

1. For a hydropower installation, the power available from water is

$$\text{Hydraulic power} = \rho g h . Q$$

where,
Q = water volume flow rate that arrives at the hydro turbine.
h = height from which water descends into the hydro turbine and is used to determine its pressure head.
ρ = water density.

Calculate the power generated with a potential hydraulic height of 80 m and a volumetric flow rate of 650 m^3/s through a hydraulic turbine with an efficiency of 0.78. Take water density as 997.75 kg/m^3. Express your answer to an appropriate number of significant digits in both watts (W) and horsepower (hp).

2. Round the following numbers to 3 significant digits.

    31.07, 4.5815, 0.002145, 4500, 7999.0

3. How many significant digits in each of the following numbers? Write each number in scientific notation.

    70.05, 1900, 0.02 × 102, 036.020, 0.000025, 8.0

4. A tire pressure gauge has a spring with a constant (k) of 300 N/m and a piston diameter of 5 mm. If the bar indicator extends 1.5 cm when the gauge is pressed against the tire valve, determine the pressure inside the tire.

5. A potentiometer is supplied with 24 Vdc and has a maximum turn angle of 300°. Calculate the output voltage if the potentiometer wiper is set at an angular position of 135°.

References

1. Norton, Harry N., 'Sensor and analyzer handbook.' Prentice-Hall, Englewood Cliffs, N.J., (1982)
2. Jacob Fraden, 'Handbook of Modern Sensors: Physics, Designs, and Applications.' 4th Edition, Springer, New York, (2010)
3. Issam A. Abu-Mahfouz, 'Instrumentation: Theory and Practice-Part II Sensors and Transducers.' Morgan and ClayPool Publishers, www.morganclaypool.com, (2022)
4. S. Schlamminger, I. Yang, and H. Kumar., 'Redefinition of SI Units and Its Implications.' MAPAN-Journal of Metrology Society of India (December 2020) 35(4): pp. 471–474. https://doi.org/10.1007/s12647-020-00421-1
5. On the Revision of the International System of Units (SI). Resolution 1. (CGPM 26th Meeting, Versailles, November 13–16, 2018), Measurement Techniques, Vol. 62, No. 5, August 2019, DOI https://doi.org/10.1007/s11018-019-01648-4
6. Bureu International des Poids et Mesures, The International System of Units (SI), 8th ed. Organisation Integouvernementale de la Convension du Metre, Paris, 2006

Dynamic Characteristics of Measuring Devices

2

2.1   Introduction

The main function of a measurement system is to infer the value of the physical variable (PV) from the sensor's output signal. For slow varying input PV, time-invariant or static conditions can be assumed when analyzing sensor sensitivity, accuracy, or when considering calibration. However, when the input PV varies at a noticeable rate, the time it takes the sensor to respond to stimulus, within its specified accuracy margin, becomes important. Sensitivity studies and design characterization of sensors focus on investigating the sensor's response in terms of magnitude (amplitude), time delay (phase shift), and shape (distortion). For dynamic measurements, an optimized sensor will produce an output that is a replica of the input signal in terms of amplitude and frequency. However, most measuring systems do not follow variations in the excitation signal with perfect conformity and do not respond instantaneously.

Such as in many engineered systems, it is common to describe the input–output relationship for a measurement device using mathematical models. These models are usually in the form of differential equations of orders and with coefficients that depend on the sensor design. Mathematical simulation of measurement systems is essential to understand their behavior and to predict the effects of the design parameters on their response to changes in the measured quantity.

This Chapter discusses the mathematical modeling of measurement systems and their response to different types of input signals. The methods discussed in this chapter can be used to model a complete measurement system or any one of its subsystems or components.

© The Author(s), under exclusive license to Springer Nature Switzerland AG 2022   13
I. Abu-Mahfouz, *Instrumentation: Theory and Practice, Part 1*, Synthesis Lectures on Mechanical Engineering, https://doi.org/10.1007/978-3-031-15246-7_2

2.2   Types of Signals

The collected values of the quantity under measurement can be presented as data with time marks designating the instants of measurements. These signals describe the response of the sensor to the measured with respect to time. Depending on the form of this response these signals may be represented by a mathematical (i.e. analytical) function. In this case the function represents the response of the system in present and future given that the system is stable and no changes in the system characteristics are expected. On the other hand, when the signals cannot be represented by an analytical function, the signals may represent a random behavior that is nondeterministic. In this case methods of statistics and probability are used to characterize the behavior of the system and to forecast the system's possible future behavior. Figures 2.1, 2.2, 2.3 and 2.4 illustrate different types of dynamic signals.

Periodic signals are those that repeat their amplitude and pattern every time cycle (period). Rotating or reciprocating machinery are known to generate signals with major periodic components under normal operating conditions. In this example, signals may represent sound, vibration, or electric power (current or voltage signals).

Combinations of periodic signals each with its own amplitude, frequency and phase my result in very complex but still stationary and periodic signals (Figs. 2.1 and 2.2). These can be decomposed into the fundamental harmonic constituents using time to frequency transform techniques such as the Fast Fourier Transform (FFT) which is discussed in Sect. 2.4.

Signal-to-noise ratio (SNR or S/N) is a measure of the quality of the signal and is defined as the ratio of the magnitude of the desired component to the magnitude of the noise component in the signal. The higher the value of the SNR the better the signal quality. Depending on the type of available data, SNR can be calculated in different ways.

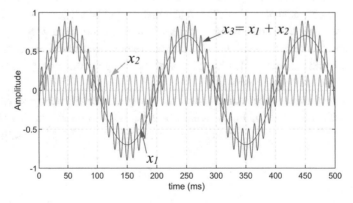

Fig. 2.1 A periodic signal $x_3(t)$ resulting from the addition of two harmonic signals, $x_1(t) = 0.7\sin(10\pi t)$ and $x_2(t) = 0.2\sin(200\pi t + 5)$

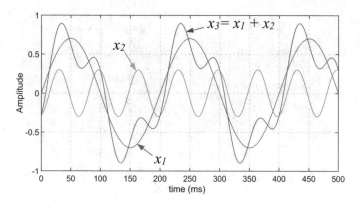

Fig. 2.2 A periodic signal $x_3(t)$ resulting from the addition of two harmonic signals, $x_1(t) = 0.7\sin(10\pi t)$ and $x_2(t) = 0.3\sin(30\pi t + 5)$

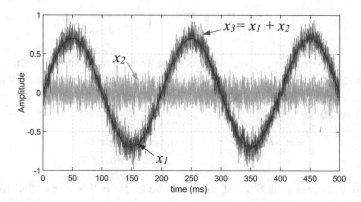

Fig. 2.3 A corrupted periodic signal $x_3(t)$ resulting from the addition of a harmonic signal $x_1(t) = 0.7\sin(10\pi t)$ and white noise signal x_2 with average amplitude of 0.1

For normal numbers (not logarithmic) SNR = S/R, divide the amplitude of the desired component by the amplitude of the noise component in the signal. Alternatively, if the measurements are expressed in decibels (logarithmic form) then SNR = S – R (subtract the noise magnitude from the desired signal magnitude). For example, a measured signal with magnitude -10 dB and noise component of -45 dB will have a SNR(dB) $= -10-(-45) = 35$ dB. However, attention should be given to the type of quantity when performing logarithmic calculations. For power (watts) data the following relation is used

$$SNR(dB) \;=\; 20\log_{10}\left(\frac{S}{N}\right)_{power} \tag{2.1}$$

whereas for data in volts the equation is

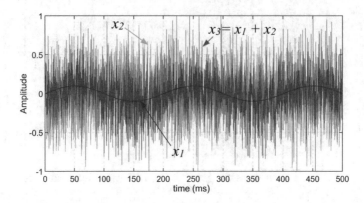

Fig. 2.4 A complex signal $x_3(t)$ resulting from the addition of two harmonic signals, $x_1(t) = 0.1\sin(10\pi t)$ and white noise signal x_2 with average amplitude of 0.3

$$SNR(dB) = 10 \log_{10}\left(\frac{S}{N}\right)_{voltage} \tag{2.2}$$

where $\left(\frac{S}{N}\right)_{power} = \left(\frac{S}{N}\right)^2_{Voltage}$ for ratios of their average respective quantities (power or voltage) using normal (not logarithmic) scale. Calculations for current are similar to the voltage calculations. For magnitudes in normal (or real) numbers the RMS (root-mean-square) values can be used

$$\left(\frac{RMS\,signal}{RMS\,noise}\right)_{voltage}, \quad \left(\frac{RMS\,signal}{RMS\,noise}\right)_{current} \tag{2.3}$$

For example, for a signal S = 50 mV and noise N = 3 mV the SNR(dB) = 10 log(0.050/0.003) = 12.2 dB.

Figure 2.5a shows a pulse signal where both h and Δt are constants. When the time interval Δt approaches zero, a special case of the pulse function called the impulse function and is depicted in Fig. 2.5b. The Dirac delta function or the unit-impulse function is an impulse function with area equal to unity. Pulse or impulse signals are observed when a system is subjected to a large force or excitation for a very short time or at one instant in time. A common example to impulse function is an impact force.

Other types of dynamic functions are shown in Fig. 2.6.

A signal that does not repeat its pattern with respect to time is cold 'aperiodic'. An example of aperiodic signal is when a deterministic signal such as a step or a ramp function occurs only once or when it occurs intermittently at different nonequal periods in time. Random or nondeterministic signals are not expressed using mathematical functions but can be characterized using statistical techniques.

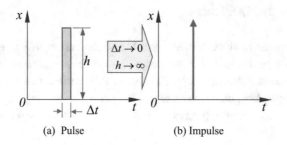

(a) Pulse   (b) Impulse

Fig. 2.5 Aperiodic pulse and impulse signals

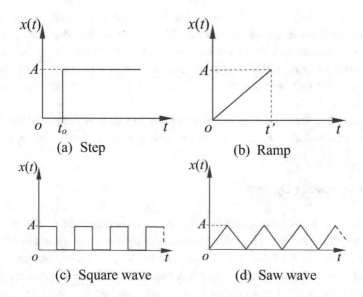

(a) Step   (b) Ramp

(c) Square wave   (d) Saw wave

Fig. 2.6 Examples of aperiodic dynamic signals: **a** $x(t) = A \ (t > t_o)$, **b** $x(t) = (A/t').t \ (0 < t < t')$, and periodic signals: **c** and **d**

2.3 Measurement Systems Models

Three generalized models of measurement systems are discussed, and these are: zeroth-order, first-order and second-order systems. The results in this discussion are universal to all systems of similar orders.

2.3.1 Zeroth-Order System

A zeroth-order sensor is characterized by a transfer function that is time independent. Such a sensor does not incorporate any energy storage devices, like a capacitor. A zeroth-order sensor responds instantaneously. In other words, such a sensor does not need any dynamic (time dependent) characteristics to be specified.

The output $x(t)$ is a scaled representation of the input stimulus $F(t)$ which is expressed as

$$x(t) = \frac{F(t)}{a_0} \tag{2.4}$$

The constant $(1/a_0)$ refers to the static sensitivity which is determined by calibration and defines the input to output relationship. This is usually the slope of the calibration curve relating the relation between the input value to the corresponding sensor output. The time parameter t shown in Eq. (2.4) is not meant to show a functional time dependency between the input and output signals but rather the instant at which the measurement would be takin.

For most real systems, the previous discussion is just an approximation that is valid for measuring systems with very fast response to a constant or a slowly varying excitation.

2.3.2 First-Order System

When measurement systems sensors do not respond instantaneously to changing stimulus or excitation. A First-Order differential equation describes a sensor or one of its instruments that incorporates a form of energy storage or energy release. Naturally, nearly any sensor will have a finite time to respond. a first-order differential equation is used to model the time dependent relationship between the dynamically changing stimulus $F(t)$ and the output signal $x(t)$ as

$$a_1 \frac{dx}{dt} + a_0 x = F(t) \tag{2.5}$$

where the coefficients a_0 and a_1 are design parameters specific to the measuring instrument. For a sudden but constant change in the input signal, Eq. (2.5). Can be rearranged in the form

$$dt = a_1 \frac{dx}{F_0 - a_0 x} \quad \Rightarrow \quad \int\limits_0^t dt = a_1 \int\limits_{x_0}^{x} \frac{dx}{F_0 - a_0 x}$$

where, $F(t) = F_0$ *for* $t > 0$ *and assuming an initial condition;*

$$x(t_0) = x_o \, at \quad t_0 = 0.$$

Solving and setting $\tau = \frac{a_1}{a_0}$, the solution to Eq. (2.5) is obtained as

$$x(t) = \underbrace{\boxed{\frac{F_o}{a_0}}}_{\text{Steady - state response}} + \underbrace{\boxed{\left(x_0 - \frac{F_0}{a_0}\right)e^{-t/\tau}}}_{\text{Transient response}} \qquad (2.6)$$

For sufficiently large time this first-order measuring system will settle near the desired value of F_0/a_0 within some margin (Error). Figure 2.7 shows a graphical presentation of the solution $x(t)$ at time t. The parameter $\tau = \frac{a_1}{a_0}$, is a universal property of the system with the dimension of time and is called the time constant. The time constant is an indication of how fast the measurement system is when responding to a change of the physical variable at its input. At one time constant the first-order system will achieve 63.21% of the full range of the input step function. Applying the standard definition of rise time as the time interval when the system arrives at 90% of the desired response (F_0/a_0) the error ratio is then

$$e^{-t_R/\tau} = 0.1, \quad or \quad t_R = 2.303\,\tau \qquad (2.7)$$

At five time constants (5τ) the error is $e^{-5} = 0.006738$ which is equivalent to the system achieving 99.32% of the desired response. For most practical applications this is good enough to assume that the system had achieved the desired steady state response.

When a harmonic excitation acts on the first-order system, the governing differential equation is expressed as

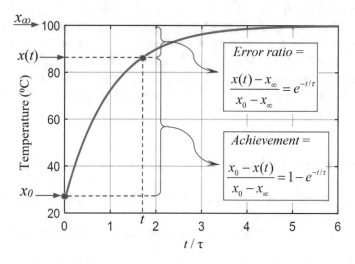

Fig. 2.7 Time response curve for the first order system

$$a_1 \frac{dx}{dt} + a_0 x = F_0 \sin \Omega t, \quad t > 0 \tag{2.8}$$

and taking as initial condition $x = x_0$ at $t = 0$, the solution to Eq. 2.8. Becomes

$$x(t) = \boxed{A . e^{-t/\tau}} + \boxed{B . \sin (\Omega t + \Phi)} \tag{2.9}$$
$$\underset{\text{Transient response}}{\qquad} \underset{\text{Steady - state response}}{\qquad}$$

where,

$$\Omega = \text{Frequency of the harmonic excitation, (rad/sec)}$$

$$\tau = \frac{a_1}{a_0} = \text{Time constant}$$

$$B = \frac{F_0/a_0}{\sqrt{1 + (\Omega \tau)^2}} \tag{2.10}$$

$$= \text{Amplitude of the steady - state response}$$

and,

$$\Phi = - \tan^{-1}(\Omega \tau)$$

$$= \text{Phase} - \text{shift angle (radians)} \tag{2.11}$$

$$\text{between input and output signals}$$

The steady-state solution is a harmonic response at the excitation frequency Ω. Figure 2.8 shows semi log plots for the relation of the steady-state response amplitude ratio $B/(F_0/a_0)$ and its phase shift Φ for wide range of the combination parameter $\Omega \tau$.

From Fig. 2.8 it can be deduced that at constant excitation frequency Ω, systems with small time constant respond with higher amplitude (stronger output signal) and with less phase lag with respect to the input stimulus. The higher the excitation frequency to be measured, the smaller the time constant must be to get an accurate measurement with no or small time delay. Amplitude ratio of unity means that the measurement system is perfectly tracking the input signal. For most practical applications, a frequency response band for which the amplitude ratio is greater than 0.707 is acceptable. This is equivalent to -3 dB $= 20 \log_{10}(0.707)$.

2.3.3 Second-Order System

Second order derivative term represents the inertia term for such as mass multiplied by acceleration in mechanical.

Generally, the mathematical model for a measurement system of the second order is defined as

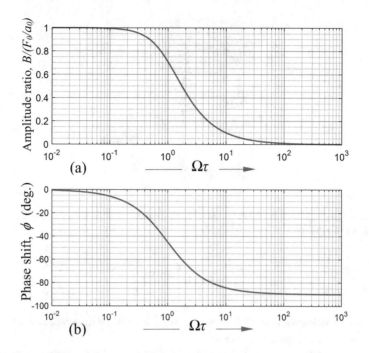

Fig. 2.8 First-order system frequency response, **a** amplitude ratio, **b** phase shift

$$a_2 \frac{d^2x}{dt^2} + a_1 \frac{dx}{dt} + a_0 x = F(t)$$

(2.12)

where a_0, a_1 and a_2 are system parameters and $F(t)$ is the excitation to the measurement system. A typical example of such second order systems is the spring-mass damper shown in Fig. 2.9. In this case $a_0 = k$ (linear spring stiffness), $a_1 = c$ (viscous damping constant) and $a_2 = m$ (system mass).

2.3.3.1 Step Excitation

Consider first the system response to a step function.

$$m \frac{d^2x}{dt^2} + c \frac{dx}{dt} + k x = F(t)$$

$$F(t) = F_0 \quad for \ t > 0$$

(2.13)

using the initial conditions and derived system parameters ζ and ω_n as defined in [1, 2],

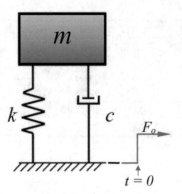

Fig. 2.9 A single-degree-of-freedom (S.D.O.F) spring mass system with a viscous damper and a step excitation of amplitude F_o

$$at \quad t = 0 \begin{cases} x_0 = 0 \\ \frac{dx}{dt} = 0 \end{cases}$$

$and\zeta = c/c_c = damping\ factor\ where,\ c_c = 2m\omega_n = critical\ damping\ constant\ and\ \omega_n = \sqrt{k/m} = natural\ undamped\ frequency$

The solution of Equation #. is presented in Table 2.1 and is plotted in Fig. 2.10. for different levels of the damping factor ζ.

The damping term $a_1 \frac{dx}{dt}$, $(a_1 = c)$ represents energy dissipation within the sensor. The response will show oscillatory transients, at a frequency ω_d, before reaching the steady-state value of F_0/k for only the underdamped case. A time constant for underdamped second order systems can be defined as $\tau = 1/\zeta\omega_n$. . In case of no damping (not realistic) the response will oscillate harmonically at the excitation frequency forever! Systems with overdamping will practically reach the steady-state value of the measured variable, within acceptable tolerance, after sufficient time.

Two quantities known as the rise time and the settling time are usually used to measure the responsivity of a dynamic system. The rise time is defined as the time it takes the system to achieve 90% of the full range from the initial state to the desired step value. The stilling time is the time required for the measurement system transients to settle within a specified % tolerance (usually ±10%) of the steady-state response. From Fig. 2.10 it can be observed that decreasing the damping factor results in shorter rise time t_R but greater overshoot and longer settling time. Measurement systems should be designed to minimize rise time, minimize overshoot, and minimize settling time. Typical damping factor range to achieve optimum performance is around $\zeta = 0.7$.

Table 2.1 Solution equations for a second-order system subject to a step function at input

Damping factor	Solution
$\zeta = 0$ Undamped	$\frac{x(t)}{F_0/k} = 1 - \cos(\omega_n t)$
$0 < \zeta < 1$ Under-damped[a]	$\frac{x(t)}{F_0/k} =$ $1 - e^{-\zeta \omega_n t}\left[\dfrac{\zeta}{\sqrt{1-\zeta^2}}\sin(\omega_d t) + \cos(\omega_d t)\right]$
$\zeta = 1$ Critically damped	$\frac{x(t)}{F_0/k} = 1 - (1 + \omega_n t)e^{-\omega_n t}$
$\zeta > 1$ Over damped	$\frac{x(t)}{F_0/k} =$ $1 - \left[\begin{array}{l} \dfrac{\zeta + \sqrt{\zeta^2 - 1}}{2\sqrt{\zeta^2 - 1}}e^{\left(-\zeta+\sqrt{\zeta^2-1}\right)\omega_n t} \\[2ex] + \dfrac{\zeta + \sqrt{\zeta^2 - 1}}{2\sqrt{\zeta^2 - 1}}e^{\left(-\zeta-\sqrt{\zeta^2-1}\right)\omega_n t} \end{array}\right]$

[a] $\omega_d = \omega_n\sqrt{1 - \zeta^2}$ is the damped natural frequency

Fig. 2.10 Response of a second-order system to a step function input at different damping levels ζ

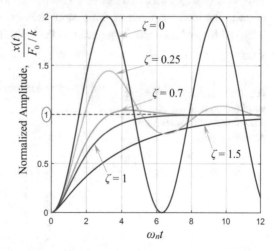

2.3.3.2 Periodic Excitation

When the second-order system of Eq. 2.12 is excited with a periodic input of the form $F(t) = F_0 \sin \Omega t$, its response will include both a transient pat and a steady state part. Ignoring the decaying transients, the steady-state response is defined as [3],

$$x(t) = \frac{(F_0/k) \cos (\Omega t - \phi)}{\sqrt{\left[1 - (\Omega/\omega_n)^2\right]^2 + [2\zeta (\Omega/\omega_n)]^2}} \tag{2.14}$$

The frequency response is a plot of the ratio of the output amplitude X_0 (when $\cos(\Omega t - \phi) = 1$) to the input amplitude F_0/k given by

$$\frac{X_0}{F_0/k} = \frac{1}{\sqrt{\left[1 - (\Omega/\omega_n)^2\right]^2 + [2\zeta (\Omega/\omega_n)]^2}} \tag{2.15}$$

And is shown in Fig. 2.11. The phase shift between the measurement system output $x(t)$ and the input stimulus $F(t)$ can be calculated using Eq. (2.16) and is shown in Fig. 2.12.

$$\phi = \tan^{-1}\left(\frac{2\zeta (\Omega/\omega_n)}{1 - (\Omega/\omega_n)^2}\right) \tag{2.16}$$

Each plot shows a family of curves with each curve representing the response at one level of damping ζ. The design of measurement system will seek to achieve a response close to the ideal response of unity amplitude ratio and zero phase-shift. This is closely achieved for underdamped systems ($\zeta \leq 0.7$) within a band of low frequency ratio values ($\Omega/\omega_n < 0.3$).

Fig. 2.11 Frequency response amplitude sensitivity plot for a second-order system with periodic input

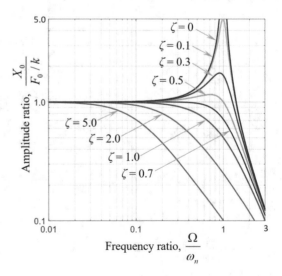

Fig. 2.12 Phase shift characteristics of the second-order system in Fig. 2.9 under periodic excitation

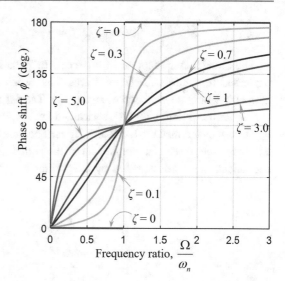

Near the resonance region $(\Omega/\omega_n) = 1$, the system response amplitude is at its peak and the phase-shift may change abruptly. Operation within the resonance region should be avoided since it may result in nonlinear distortions and possible damage to some measurement systems. It is also observed from Fig. 2.11 that the amplitude of the measured signal is attenuated (filtered out) at large frequencies.

The phase shift plot (Fig. 2.12) shows nonlinear relations for almost all frequency ratio values. However, at damping level of $\zeta = 0.7$ the relation between phase shift and the frequency is closer to linear behavior than for the other levels of damping. This is yet another reason to have measurement systems design with damping ratio close to 0.7. These characteristics highlights the importance of using modeling to study measurement system design. They also provide guidance to match the selection of sensors to the desired signal characteristics to be measured.

2.4   Signals in Time and Frequency Domains

Measurements usually consist of the magnitude (amplitude) of the physical variable as a function of time. Each measurement is usually made of several constituent components (signals) each with its own amplitude and frequency. Any periodic function y(t) may be represented by a trigonometric seires of the form [4, 5];

$$y(t) = a_0 + \sum_{n=1}^{\infty} a_n \cos(\omega_n x) + b_n \sin(\omega_n x) \qquad (2.17)$$

where, $\omega_n = n\omega$

$$\omega = 2\pi f = \frac{2\pi}{T}, \quad and \quad f_n = nf$$

The circular frequency ω is expressed in radians/sec and the related frequency f is expressed in cycles/sec, Hz, or Hertz. The series index n is used to indicate the nth harmonic within the function or signal $y(t)$. The fist harmonic ($n = 1$) is called the fundamental term or harmonic in the series and has the lowest frequency ω or f. The first term a_0 represents the mean value of the series. The constants a_n and b_n are called the Fourier coefficients represent the amplitudes of the components (harmonics) in the signal. These coefficients can be determined by integrating Eq. (2.17) over one period T of the series $y(t)$.

$$a_0 = \frac{1}{T} \int_{-T/2}^{T/2} y(t)dt$$

$$a_n = \frac{2}{T} \int_{-T/2}^{T/2} y(t)\cos\left(\frac{2\pi nt}{T}\right)dt$$

$$b_n = \frac{2}{T} \int_{-T/2}^{T/2} y(t)\sin\left(\frac{2\pi nt}{T}\right)dt \tag{2.18}$$

A periodic function y(t) that is even, $y(t) = y(-t)$, is represented by only cosine terms

$$y(t) = a_0 + \sum_{n=1}^{\infty} a_n \cos(2\pi nft) = \sum_{n=0}^{\infty} a_n \cos(2\pi nft) \tag{2.19}$$

While an odd function, $y(t) = -y(-t)$, is represented by only sine terms

$$y(t) = \sum_{n=1}^{\infty} b_n \sin(2\pi nft) \tag{2.20}$$

A function that is neither even nor odd is represented by a Fouries series containing both sine and cosine terms, Eq. (2.17). By introducing the concept of the phase angle, the Fourier series in Eq. (2.17) may be presented as

$$y(t) = a_0 + \sum_{n=1}^{\infty} c_n \cos(2\pi f_n t - \phi_n) \tag{2.21}$$

where the amplitude c_n is determined by

$$c_n = \sqrt{a_n^2 + b_n^2} \tag{2.22}$$

Fig. 2.13 A square wave with amplitude 1 and peak-to-peak amplitude 2

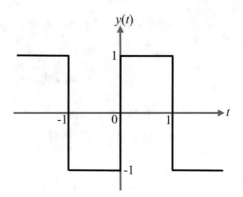

And the phase angle ϕ_n is determined by

$$\phi_n = \tan^{-1}\left(\frac{b_n}{a_n}\right) \tag{2.23}$$

As an example, Let's determine the Fouries series terms that represent a square wave shown in Fig. 2.13. The wave function for once cycle $(-1 < t < 1)$ can be describe by

$$y(t) = \begin{cases} 1 & 0 < t < 1 \\ -1 & -1 < t < 0 \end{cases}, \quad \text{and} \quad T = 2. \tag{2.24}$$

This is an odd function and therefore its Fourier series contain only the sine terms (Eq. (2.20)) and $a_0 = 0$.

$$y(t) = \sum_{n=1}^{\infty} b_n \sin(2\pi n f t)$$

The amplitudes b_n are obtained by

$$b_n = \frac{2}{T} \int_{-T/2}^{T/2} y(t) \sin\left(\frac{2n\pi t}{T}\right) dt, \quad \text{where } T = 2$$

$$b_n = \int_{-1}^{0} (-1) \sin\left(\frac{2n\pi t}{2}\right) dt + \int_{0}^{1} (1) \sin\left(\frac{2n\pi t}{2}\right) dt$$

$$b_n = \left[\frac{1}{n\pi} \cos(n\pi t)\right]_{-1}^{0} + \left[\frac{-1}{n\pi} \cos(n\pi t)\right]_{0}^{1}$$

$$b_n = \frac{1}{n\pi}(1 - \cos(-n\pi) - \cos(n\pi) + 1)$$

It can be verified that for even values of $n \to b_n = 0$, and that for odd values of $n \to b_n = 4/n\pi$.

$$
\left.\begin{array}{l}
y(t) = \sum_{n=1}^{\infty} b_n \sin(2n\pi f t) = \frac{4}{\pi} \sum_{n=1}^{\infty} \frac{1}{n} \sin(2n\pi f t), \\[2mm]
\text{where } n = 1, 3, 5, 7, \ldots \\[1mm]
\text{and } f = 1/T = 1/2 \\[1mm]
Thus, \\[1mm]
y(t) = \frac{4}{\pi} \sum_{n=1}^{\infty} \frac{1}{n} \sin(n\pi t) \\[2mm]
y(t) = \frac{4}{\pi} \sin(\pi t) + \frac{4}{3\pi} \sin(3\pi t) + \frac{4}{5\pi} \sin(5\pi t) + \ldots
\end{array}\right\} \qquad (2.25)
$$

The fundamental harmonic ($n = 1$) is

$$
\omega = 2\pi f = \frac{2\pi}{T} = \frac{2\pi}{2} = \pi
$$

Higher harmonics are 3π, 5π, 7π. Figure 2.14 shows the summation of only the first seven odd harmonics as presented in (Eq. (2.25)). As more harmonics are added to the summation, the result becomes more smoother and closer to a square wave which is shown in red in Fig. 2.14.

Measurement data for a variable $y(t)$ is practically sampled, acquired, and recorded in a discrete or digital form. Sampling and data acquisition concepts are discussed in Chapter 6. A measurement $y(t)$ sampled at a rate (f_s samples/sec) for an acquisition period T (seconds) will consist of N data points separated by Δt equal intervals.

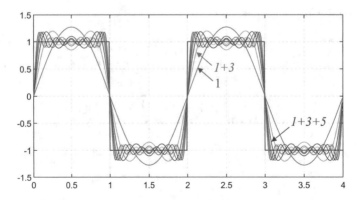

Fig. 2.14 The square wave and the first seven sums of its Fourier series harmonics as presented by Eq. (2.25)

$$N = f_s T = \frac{T}{\Delta t}$$

Each sample $y(t_i)$ is sampled at time instant t_i,

$$t_i = i\,\Delta t \quad \text{for} \quad i = 0, 1, 2, \ldots, N-1$$

The Fourier series coefficients in Eq. (2.17) are obtained using computer numerical integration algorithms and can be presented by the following summations for N samples [6].

$$a_0 = \frac{1}{N} \sum_{i=0}^{N-1} y(i\,\Delta t)$$

$$a_n = \frac{2}{N} \sum_{i=0}^{N-1} y(i\,\Delta t) \cos\left(\frac{2\pi i n}{N}\right)$$

$$b_n = \frac{2}{N} \sum_{i=0}^{N-1} y(i\,\Delta t) \sin\left(\frac{2\pi i n}{N}\right)$$

$$\text{for} \quad n = 1, 2, \ldots, (N/2 - 1) \tag{2.26}$$

where a_0 represents the mean or average value of the measured discrete samples. Similar to the continuous Fourier's series, Eq. (2.26) presents the coefficients for the discrete Fourier series transform (DFT) which creates a series of components each with its own frequency ($f_i = n/N\Delta t$) and amplitude from the original discrete time measurements. This decomposition of the time domain into components that can be reconstructed (transformed) into the frequency domain is accomplished by software that is embedded with the data acquisition device, as a stand alone package, or as algorithms within platform packages such as MATLAB®. The fast Fourier transform (FFT) is a fast algorithm [7] that calculates the coefficients and harmonics of the DFT at high speed.

2.5 Examples

Example 2.1 A first-order temperature measuring device initially at room temperature (25 °C) is suddenly immersed in hot water at 85 °C. If the time constant for this sensor is 13 s,

(a) Determine the rise time for its reponse.
(b) Determine the temperature when the sensor is 2% in error.

Solution

(a) The rise time t_R is the time it takes the sensor to achieve 90% of the full range of measurement. From Eq. (2.7)

$$e^{-t_R/\tau} = 0.1, \quad or \quad t_R = 2.303\,\tau$$
$$t_R = 2.303 \times 13\ s = 29.94\ s$$

(b) The error ratio is 0.02 and using the appropriate relation from Fig. 2.7

$$e^{-t/\tau} = 0.02 \Rightarrow \frac{T(?) - T_\infty}{T_o - T_\infty} = \frac{T(?) - 85}{25 - 85} = 0.02$$
$$\Rightarrow T(?) = 83.8\ ^\circ C$$

Example 2.2 Figure 2.15 shows a response curve for a second-order measurement system subjected to a unit step excitation starting at time $t = 0$. Use this curve to estimate the damped natural frequency (ringing frequency), the rise time t_R, the settling time to within 2% of the final steady state response. If the damping factor is estimated to be $\zeta = 0.2$, determine the natural undamped frequency ω_n of the system.

Solution
The damped period between first two peaks $T_d = (0.92 - 0.27) = 0.65$ s which corresponds to a damped natural frequency of $\omega_d = 1/0.65 = 1.539$ Hz $= 9.67$ rad/s.

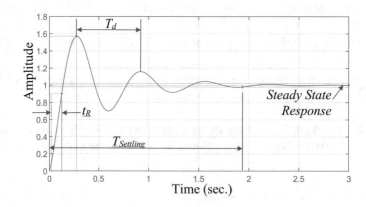

Fig. 2.15 Response curve for a second-order system to a unit-step input

The rise time is the time when the response is between 10% and 90% of the full range. For a unit step function this the time to achieve from 0.1 to 0.9 and is estimated from the curve as $t_R = 0.11$ s.

The settling time is measured from start of response to the first instant when the response is limited to within 2% of the final value and is estimated as $T_{Settling} = 1.95$ s.

The relation between the damped and undamped natural frequencies is

$$\omega_d = \omega_n\sqrt{1-\zeta^2} \rightarrow \omega_n = \frac{\omega_d}{\sqrt{1-\zeta^2}} = \frac{1.539\,\text{Hz}}{\sqrt{1-(0.2)^2}}$$

$$\omega_n = 1.571\,\text{Hz} = 9.88\,\text{rad/s}.$$

Example 2.3 Figure 2.16 displays three sine waves each with it own frequency and amplitude. The vertical ordinate scale is kept fixed for easy visual comparison between the signals amplitudes. Plot the signal resulting from addition of these three harmonics and use the FFT algorithm to plot the frequency spectrum of the resulting signal.

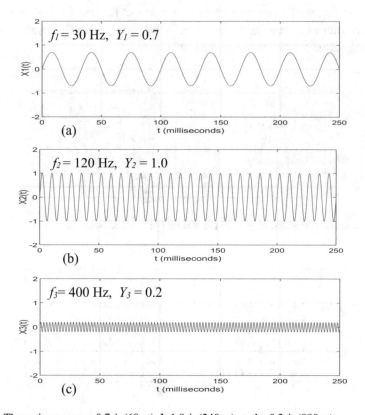

Fig. 2.16 Three sine waves **a** 0.7sin(60πt), **b** 1.0sin(240πt), and **c** 0.2sin(800πt)

Solution

The result of adding the three signals shown in Fig. 2.2 is depicted in Fig. 2.2. If this represents a measured variable, it would be difficult to infer the details of the constituents from the time domain alone. Using the FFT enables the extraction of the amplitude and frequency details for the three constituent harmonic waves as clearly illustrated in the amplitude spectrum shown in Fig. 2.2.

To further the discussion in this example, a random signal (white noise) shown in Fig. 2.19a is added to the signal in Fig. 2.17 and the result is presented in Fig. 2.19. The FFT spectrum for this signal (Fig. 2.20) still shows the three original harmonics but also shows a low amplitude broadly spread spectrum resulting from the white noise.

If the amplitude of the white noise component in the signal in Fig. 2.19 is increased to 0.2 instead of 0.075 the resulting FFT spectrum (Fig. 2.21) depicts only the first two harmonics with larger amplitudes. The third harmonic component with amplitude equal to that of the white noise is not observed since it is completely corrupted with the equal strength (amplitude) random signal.

Example 2.4 Plot the Fourier time series summation of the first five odd harmonics of a square wave with a fundamental frequency of 20 Hz and use the FFT algorithm to plot the frequency spectrum for the resulting time signal.

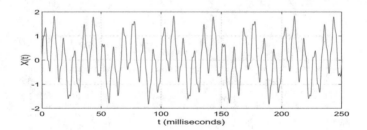

Fig. 2.17 A period signal resulting from the addition of the three sine waves in Fig. 2.16

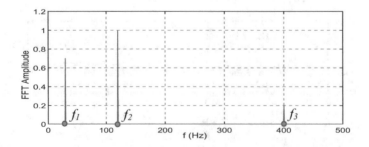

Fig. 2.18 Amplitude spectrum for the signal in Fig. 2.17 obtained using a FFT algorithm (Fig. 2.18)

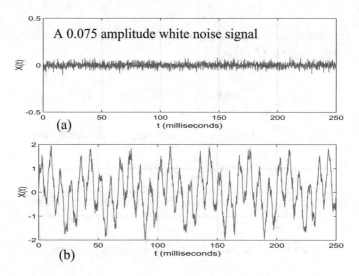

Fig. 2.19 **a** Random signal, **b** result of adding the noise to the signal in Fig. 2.17

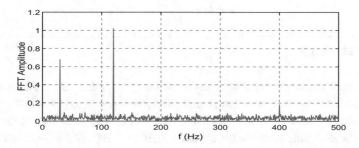

Fig. 2.20 Amplitude spectrum for the signal in Fig. 2.19 obtained using a FFT algorithm

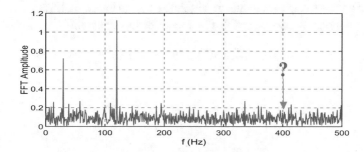

Fig. 2.21 FFT spectrum with an added 0.2 amplitude white noise

Solution

The summation of the first five odd harmonics in the Fourier series representing the square wave function given by Eq. (2.25) is

$$y(t) = \frac{4}{\pi} \sum_{n=1}^{9} \frac{1}{n} \sin(n\pi t), \quad n = 1, 3, 5, 7, and\ 9$$

$$y(t) = \frac{4}{\pi} \sin(40\pi t) + \frac{4}{3\pi} \sin(120\pi t) + \frac{4}{5\pi} \sin(200\pi t)$$

$$+ \frac{4}{7\pi} \sin(280\pi t) + \frac{4}{9\pi} \sin(360\pi t)$$

In this form the first five odd harmonics (in Hz) are; 20, 60, 100, 140, and 180. In radians these are;

$$\omega = 2\pi f \Rightarrow 40\pi, \ 120\pi, \ 200\pi, \ 280\pi, \text{ and } 360\pi.$$

The time series is depicted in Fig. 2.22a and the corresponding FFT spectrum is shown in Fig. 2.22b.

2.6   Problems

2.1 A temperature transducer that is designed based on a first-order model is used to measure a slowly fluctuating temperature at a frequency of 0.02 Hz. If the transducer has a time constant of 10 s, determine the time delay (phase shift) and the reduction in amplitude during its harmonic response. (Hint, use Eqs. (2.10, 2.11), and Fig. 2.8).

2.2 The rise time t_R for a first-order measuring system is experimentally determined to as 37 s. What is the instant in time at which this system response can be accepted as the steady state response within an absolute error tolerance of 1.0%?

2.3 A second-order system is modeled by the following equation

$$0.01\ddot{x} + 2.5\dot{x} + 15000x = 0.3\sin(200t)$$

Where,

$$\ddot{x} = \frac{d^2x}{dt^2}, \quad and \quad \dot{x} = \frac{dx}{dt}$$

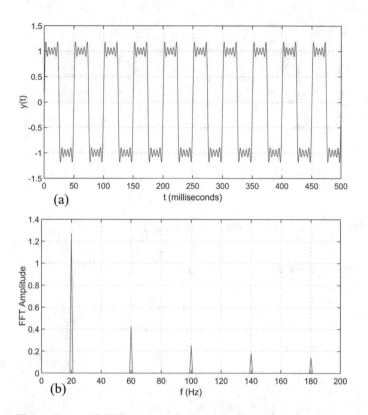

Fig. 2.22 **a** Time series, and **b** FFT spectrum for Example (2.4) solution

Calculate the following system parameters.

(a) Mass, m (kg)
(b) Undamped natural frequency, ω_n (Hz)
(c) Damping factor, ζ
(d) Damped natural frequency, ss ω_d (Hz)
(e) Excitation frequency, Ω (Hz)
(f) Phase shift, ϕ (deg.)

Determine and plot the frequency response of a second-order system that has a mass of 0.05 kg, a damping factor $\zeta = 0.63$, and supported by a stiff membrane with spring constant $k = 120$ kN/m. What is the range of the frequency ratio for which the system sensitivity (amplitude ratio) is equal to 1?

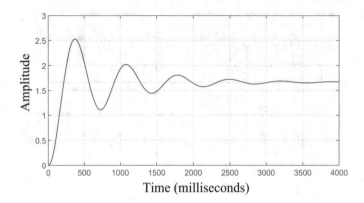

Fig. 2.23 Time response to a step function for Problem 2.5

Fig. 2.24 Wave functions for Problem 2.6

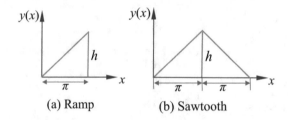

(a) Ramp   (b) Sawtooth

Figure 2.23 shows a response curve for a second-order measurement system subjected to a step excitation starting at time $t = 0$. Use this curve to estimate the damped natural frequency (ringing frequency), the rise time t_R, the steady state response (measurement value), and the settling time to within 2% of the final steady state response.

Determine the Fourier series for each of the wave functions shown in Fig. 2.24.

References

1. Charles M. Close and Dean K. Frederick and Jonathan C. Newell, Modeling and analysis of dynamic systems, 3rd ed., Wiley, New York, 2002.
2. Katsuhiko Ogata, System Dynamics, 4th ed., Pearson, Prentice-Hall, Upper Saddle River, NJ, 2004.
3. William J. Palm III, System dynamics, 3rd edition. McGraw-Hill Science, New York, NY, 2014.
4. Hanna, J. Ray and John H. Rowland. Fourier series, transforms, and boundary value problems. 2nd ed., New York : Wiley, 1990.

5. Brecewell, Ronald. N. The Fourier Transform and its Applications, 3rd ed., Boston : McGraw Hill, 2000.
6. Brown, J.S., and R.V. Churchill; Fourier Series and Boundary Value Problems, 8th ed, McGraw-Hill, 2011.
7. Cooley, J. W., and J. W. Tukey; An algorithm for the Machine Calculation of Complex Fourier Series. Mathematics of Computation, 1965, v. 19, no. 90, pp:249–259.

Data Analysis

3

3.1 Introduction

The objective of any measurement system is to provide a reading that presents, as accurate as possible, the true value of the quantity under measurement. For most engineering applications this requires taking several measurements and performing analysis on the complete set of collected data or on a subset of the data and come up with a single quantity that represents the average or mean value of the measurements. Next it is important to find out how accurate is the average quantity in representing the true value of the physical variable (PV). Variations of the measured data need also be characterized relative to the intended application. Sources of variations in the measurements and the limits of these variations are also to be defined. The theory and methods of probability and statistics will assist the experimentalist in data interpretation, error estimation, and pattern recognition in the measured data.

3.2 Statistical Analysis

Sources of noise and variations will cause the data to be scattered or distributed in different forms within the domain of interest. These distributions may typically show combinations of uniform patterns and random scatter. Even for data collected under controlled test conditions, the data points will take values at or about a specific value more frequently than other values. This value of high frequency (tendency) of occurrence is called the central tendency point of the distribution. This central tendency point is expected to best represent the true average (*mean*) value of the measurements. Table 3.1 shows a sample

© The Author(s), under exclusive license to Springer Nature Switzerland AG 2022
I. Abu-Mahfouz, *Instrumentation: Theory and Practice, Part 1*, Synthesis Lectures on Mechanical Engineering, https://doi.org/10.1007/978-3-031-15246-7_3

of a measured variable taken randomly at different instances in time but under similar test conditions. Figure 3.1 is a plot of these measured values in which the data is arranged into groups or intervals. The height of each bar in the histogram represents the number of data points whose values occur within the interval represented by that bar. It is clear that more data points cluster about the mean. A non-dimensional quantity for the height of each bar in Fig. 3.1, called the frequency distribution, can be defined for the jth interval as:

$$fd_j = \frac{Number\,of\,data\,points\,in\,interval\,j}{Total\,number\,of\,measurements} = \frac{n_j}{N} \tag{3.1}$$

where,

$$N = \sum_{j=1}^{M} n_j, \quad and \quad \sum_{j=1}^{M} fd_j = 1.$$

M is the total number of subdivided intervals between the lower limit (LL) and the upper limit (UL) of the measured values of the PV. We can say that the probability of a measured value to occur within an interval j is represented by the frequency distribution fd_j. As M increases, the interval size $\delta_j = \frac{UL-LL}{M}$ decreases and on the limits $N \to \infty$ and $\delta_j \to 0$ the frequency distribution for a PV measurement to assume a particular value x can be represented by the probability density function $p_d(x)$.

Table 3.1 Sample data from measuring a constant force of 100 N applied by a pneumatic cylinder during a controlled experiment

j	Force interval (N)	n_j	$fd_j = n_j/Na$
1	$99.9 \leq x_i < 100.0$	1	0.001
2	$100.0 \leq x_i < 100.1$	3	0.003
3	$100.1 \leq x_i < 100.2$	16	0.016
4	$100.2 \leq x_i < 100.3$	61	0.061
5	$100.3 \leq x_i < 100.4$	147	0.147
6	$100.4 \leq x_i < 100.5$	235	0.235
7	$100.5 \leq x_i < 100.6$	257	0.257
8	$100.6 \leq x_i < 100.7$	172	0.172
9	$100.7 \leq x_i < 100.8$	77	0.077
10	$100.8 \leq x_i < 100.9$	25	0.025
11	$100.9 \leq x_i < 101.0$	4	0.004
12	$101.0 \leq x_i < 101.1$	2	0.002

[a] $N = 1000$ data points

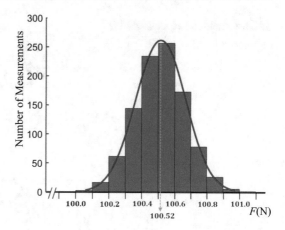

Fig. 3.1 Histogram ($\Delta F =$ 0.1 N) representing the data in Table 3.1

In Fig. 3.1 the $p_d(x)$ for the data in Table 3.1 is approximated by the red curve fitting the peak points of the bars in the histogram. This curve looks similar to a normal (or Gaussian) distribution.

There are several standard probability density distributions that are usually followed by measured data in most of engineering and scientific applications [1–3]. Once a standard distribution function for a particular set of measurements is determined, then it becomes easier to extract several useful parameters and features that will assist in data interpretation. The following equation can be used to determine the true average (mean) value of a randomly measured variable x,

$$\mu = \lim_{N \to \infty} \frac{1}{N} \sum_{i=1}^{N} x_i \tag{3.2}$$

The true variance indicates the width of the distribution of the randomly measured variable and it is given by,

$$\sigma^2 = \lim_{N \to \infty} \frac{1}{N} \sum_{i=1}^{N} (x_i - \mu)^2 \tag{3.3}$$

The true standard deviation (a commonly used statistical parameter) is related to the width or spread of the data distribution function and is the square root of the variance,

$$\sigma = \lim_{N \to \infty} \sqrt{\frac{1}{N} \sum_{i=1}^{N} (x_i - \mu)^2} \tag{3.4}$$

Equations (3.2–3.4) assume infinite number of data points ($N \to \infty$). For a finite size sample of random measurements N of the variable x, estimates of the sample mean, sample variance and sample standard deviation are representatives only of their sample N

and are defined respectively by

$$\mu_s = \frac{1}{N} \sum_{i=1}^{N} x_i, \tag{3.5}$$

$$\sigma_s^2 = \frac{1}{(N-1)} \sum_{i=1}^{N} (x_i - \mu_s)^2, \tag{3.6}$$

and

$$\sigma_s = \sqrt{\frac{1}{(N-1)} \sum_{i=1}^{N} (x_i - \mu_s)^2} \tag{3.7}$$

For the sample variance σ_s^2 and the sample standard deviation σ_s, $(N-1)$ is used for the degrees of freedom instead of N (in case of infinite statistics) because the previously determined sample mean μ_s is used in these equations. These sample parameters are only approximations of the true parameters which can only be theoretically obtained from infinite (very large) data sets. The range $(x_i - \mu_s)$ is the error or the deviation (d_i) of the measured variable (data point x_i) from the sample mean μ_s.

3.3   The Normal (Gaussian) Distribution

This is probably the most important and most used probability density function in the field of statistics. Many statistical analysis techniques assume that measurements data in engineering and other fields of science follow a normal distribution.

The normal or Gaussian distribution is also called the 'bell-curve' distribution because of its shape, where the data is symmetrically distributed about the mean as shown in Fig. 3.2. The analytical formula for the probability density function (pdf) for a random variable x that follows the normal distribution is:

$$p_d(x) = \frac{e^{\left[\frac{-(x-\mu)^2}{2\sigma^2}\right]}}{\sigma\sqrt{2\pi}} \tag{3.8}$$

where μ is the central tendency of the distribution located at the mean and σ^2 is the variance which serves also as a scaling parameter for the normal distribution. Large values of σ will stretch wider the distribution curve (i.e., data is more broadly distributed about the mean). Small values of σ will compress or squeeze the curve closer around the mean while increasing the maximum value of the distribution function $p_d(x)$. Figure 3.3 shows the effect of changing both parameters μ and σ on the normal distribution. When $\mu = 0$ and $\sigma = 1$ the $p_d(x)$ is called the *standard normal distribution* and it is given by

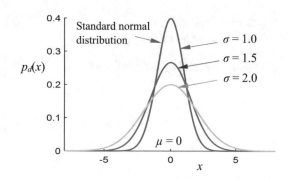

Fig. 3.2 Standard normal distribution for three values of the standard deviation

$$p_d(x) = \frac{e^{\left(-\frac{x^2}{2}\right)}}{\sqrt{2\pi}} \tag{3.9}$$

In this discussion it is assumed that no systematic error is present in the data and therefore the mean value μ is the most probable representation of the true value of the measurement. Knowing the analytical expression of a probability distribution helps in predicting the probability that a measured value x will fall within any defined interval. This probability is equal to the area under the $p_d(x)$ curve bounded by the lower limit x_{LL} and upper limit x_{UL} of the interval of interest as illustrated in Fig. 3.4 and is given by

$$p_d\{x_{LL} \leq x \leq x_{UL}\} = \int_{x_{LL}}^{x_{UL}} p_d(x)\mathrm{d}x \tag{3.10}$$

For most measurements we are interested more in the interval (error range) symmetrically surrounding the expected true mean μ value of the random variable x such that x will fall within the range $\mu \pm \beta\sigma$ or, $\{\mu - \beta\sigma \leq x \leq \mu + \beta\sigma\}$. Making use of the following change of variables,

Fig. 3.3 The effect of changing μ and σ on the normal distribution

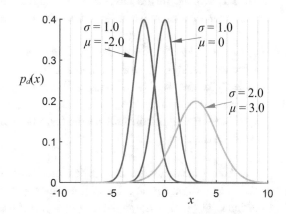

Fig. 3.4 The probability that x will fall within an interval is equal to the area under the $p_d(x)$ curve bounded by the interval

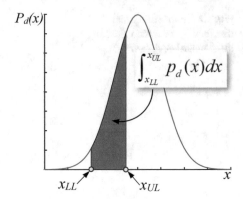

$$\beta = \frac{(x - \mu)}{\sigma}, \Rightarrow dx = \sigma d\beta \tag{3.11}$$

and for any interval limit x_i, z_i will be used to define the limit in the standard form,

$$z_i = \frac{(x_i - \mu)}{\sigma} \tag{3.12}$$

the normal density function (Eq. 3.8) can be simply put into the standard normal density function form.

The standard normal error function $p_d(\beta)$ is

$$p_d(\beta) = \frac{1}{\sqrt{2\pi}} \int_{-z_i}^{z_i} e^{\left(-\frac{\beta^2}{2}\right)} d\beta = \frac{2}{\sqrt{2\pi}} \int_{0}^{z_i} e^{\left(-\frac{\beta^2}{2}\right)} d\beta \tag{3.13}$$

where β can be defined as the standardized normal variable for any random variable x.

As illustrated in Fig. 3.5 we can predict the probability that a measurement of the variable x_i will lie in the interval defined by the limits $\mu \pm z_i\sigma$. The following three estimates are of interest when using the normal error function.

$z_i = 1$, $68.26\% \approx 68\%$ of data (i.e., area under the $p_d(x)$ curve) fall within the first standard deviation ($\pm \sigma$) from the mean μ.

$z_i = 2$, $95.45\% \approx 95\%$ of data fall within two standard deviations ($\pm 2\sigma$) from the mean μ.

$z_i = 3$, 99.7% (approximately all the data) fall within three standard deviations ($\pm 3\sigma$) from the mean μ.

These three statements are also called the 'Empirical Rule' or the '68-95-99.7 Rule' or the 'Three Sigma Rule'. These fundamental statistical concepts can be readily used to estimate or predict the probability of a measured variable having a specific value or

Fig. 3.5 Standard deviation
characterization of the
probability density function for
a Gaussian distribution

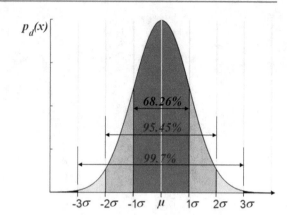

accruing within a certain interval. The probability that a variable x_i will lie between
$\mu \pm z_i \sigma$ is

$2 \times p_d(z_i)$. The sample mean μ_s for a random sample of size N is expected to approxi-
mate the true mean μ of the population within some $\pm$ *confidence interval* that depends on
the sample variance. Taking M randomly selected sample sets each with N measurements
from a large population of measurements for the same variable x will yield M random
estimates of sample means $\mu_{sj}, j = 1, 2,..., M$. This discrepancy is due mainly to the random errors associated with the time of measurements in addition to other possible random
errors. These M values of the samples means will follow a normal distribution about the
population (true) mean. Also, the standard deviation of the sample means distribution is
related to that of the population by the relation

$$\sigma_{\mu_s} = \frac{\sigma}{\sqrt{N}} \text{ (standard deviation)}$$

$$\sigma_{\mu_s}^2 = \frac{\sigma^2}{N} \text{ (variance)} \tag{3.14}$$

This indicates that the larger is the sample size N, the smaller is the variance and
thus the closer are the samples means μ_{sj} distributed about the population true mean μ.
According to the central limit theorem, as the sample size N becomes large, the sampling
distribution of the mean approaches the normal distribution and becomes centered at the
true population mean, μ.

3.4   The Student's *t*-Distribution

Developed by William S. Gosset, the Student's t distribution is attempted to produce sat-
isfactory prediction results when using small sample size or finite samples with unknown

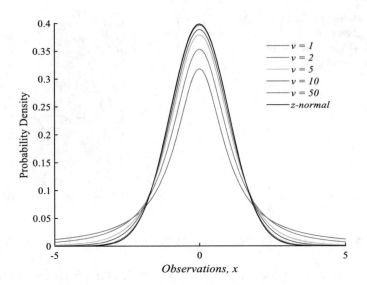

Fig. 3.6 is a family of curves depending on a single parameter ν (the degrees of freedom). As the degrees of freedom ν approach infinity, the t distribution approaches the standard z normal distribution

population standard deviation σ. The sample standard deviation σ_s is used in the formulation of the t distribution. Since there are more uncertainties in relying on small data sets, the t distribution shows more dispersion than the normal distribution as shown in Fig. 3.6. However, as the sample size increases, the t distribution gradually approaches the z (normal) distribution. Therefore, for large samples ($\nu > 50$) it is recommended to use the z distribution (Fig. 3.6). The t distribution is very similar to the standard normal z distribution and hence the t variable can be expressed as given by Eq. (3.15).

For the same probability $p_d\%$ (area under the distribution curve) the t distribution gives a wider estimate for the confidence interval than the standard normal error z distribution. Also, unlike the z distribution, estimates of confidence intervals for the t distribution are not fixed and are functions of the sample size or ν (i.e., ν is the sample degrees of freedom $\nu = N-1$).

$$z = \frac{\mu_s - \mu}{\sigma/\sqrt{N}} \qquad \underset{N \to \infty}{\longleftarrow} \qquad t_\nu = \frac{\mu_s - \mu}{\sigma_s/\sqrt{N}} \qquad (3.15)$$

$$\underset{\substack{\text{Constant statistical variables} \\ \text{(usually unknown)}}}{} \qquad\qquad \underset{\substack{\text{Sample dependent} \\ \text{statistical variables}}}{}$$

Given a probability level $p_d\%$, the confidence interval (centered about the sample mean μ_s) within which the true mean value μ might fall is presented by the $\pm\left(t_{\nu,p_d}\sigma_s\right)$ quantity in the following equation.

$$\mu = \mu_s \pm \left(t_{v,p_d}\sigma_s\right) = \mu_s \pm \left(t_{v,p_d}\frac{\sigma}{\sqrt{N}}\right) \tag{3.16}$$

Equation (3.16) is an estimate of the population true mean value based on the statistics of a finite data set of the measured variable x. We can accept that for 95% confidence interval the sample mean will fall in the range $\pm 2\sigma_s$.

3.5 Chi-Square (χ^2) Distribution and the Goodness-of-Fit Test

It is sometimes of significant importance to test how close the spread of a finite size sampled data can represent the true behavior of the measured variable if large data could be available. In many practical applications it is not always feasible or even possible to collect sufficiently large data and some assumptions or (common sense) decisions based on finite data sets are to be made. This is a type of statistical inference in which we use one or more finite data sets to estimate the parameters of the underlying distribution of the measured values within a specified confidence interval. One of the parameters that the Chi-square (χ^2) distribution assist in evaluating is the variance which is an indication of the data scatter about the mean (i.e., a deviation or error parameter). From Eqs. (3.6) and (3.12) we can write

$$\sigma_s^2 = \frac{1}{(N-1)}\sum_{i=1}^{N}(x_i - \mu_s)^2 = \frac{1}{v}\sum_{i=1}^{N}(x_i - \mu_s)^2 \tag{3.17}$$

$$z_i = \frac{(x_i - \mu)}{\sigma} \Rightarrow z_i^2 = \frac{(x_i - \mu)^2}{\sigma^2} \tag{3.18}$$

The χ^2 distribution is a distribution of the sum of squared standard normal errors resulting from n data sets each with N data points taken from the same population. Using Eqs. (3.17) and (3.18) and assuming a normal distribution for each data set, the χ^2 is given by the general form

$$\chi^2 = \frac{v\sigma_s^2}{\sigma^2}, \quad \text{or} \tag{3.19a}$$

$$\chi^2 \approx z_1^2 + z_2^2 + \cdots + z_n^2 \tag{3.19b}$$

The degrees of freedom v is defined as the independent number of data sets. Figure 3.7 shows that the χ^2 distributions are represented by nonsymmetrical (skewed) curves with the curve shape depending on v. It can also be clearly noticed that as the degrees of freedom increases the distribution curve approaches that of the normal distribution. Just like other probability distribution curves, the total area under the χ^2 distribution curve is equal to unity (Fig. 3.8). The area under the $P(\chi^2)$ curve between any two chi-square values is equal to the probability that the value of a χ^2 statistic will fall between these two limits. Therefor the probability that the sample variance $\sigma_s{}^2$ estimates the population or true variance σ^2 with a level of significance α (or with a level of confidence $1-\alpha$) can be determined by

$$P\left\{\chi^2_{1-\frac{\alpha}{2}} \le \chi^2 \le \chi^2_{\frac{\alpha}{2}}\right\} = 1 - \alpha \tag{3.20}$$

Or,

$$P\left\{\frac{v\sigma_s^2}{\chi^2_{\frac{\alpha}{2}}} \le \sigma^2 \le \frac{v\sigma_s^2}{\chi^2_{1-\frac{\alpha}{2}}}\right\} = 1 - \alpha \tag{3.21}$$

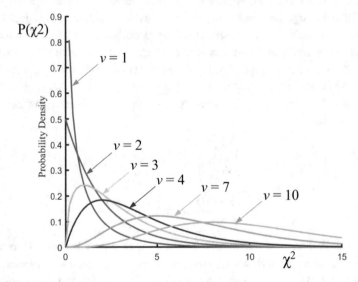

Fig. 3.7 χ^2 distribution for several degrees of freedom v

Fig. 3.8 The χ^2 distribution and the level of significance α

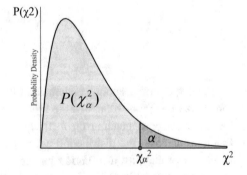

The **Goodness-of-fit test** uses the χ^2 statistics to quantify the discrepancy between the variation of the randomly sampled measured variable and the variation of an assumed (hypothetical) distribution function. The following quantity is usually used together with the degree of freedom to calculate (using software) or to read using tables the $P()$ values that will assist users with statistical inference and data interpretation.

$$\chi^2 = \sum_{j}^{M} \frac{\left(n_{j,measured} - n_{j,expected}\right)^2}{n_{j,expected}}, \quad j = 1, 2, ..., M \tag{3.22}$$

where M is the number of intervals of measurements, or data categories, within the N data points and each jth interval contains n_j measurement values of the variable x. A rule of thumb, resulting from the *central limit theorem*, is that n_j should be at least 5. The values $n_{j,expected}$ are the assumed or theorized values with an expected distribution of interest. The degrees of freedom in the measured intervals or groups $v = M - k$, where k is the number of constraints or conditions imposed on the data and or the assumed distribution. For a value of $\chi^2 = 0$, the measured distribution will match the expected (assumed) distribution which is '*too good to be true!*'. The greater the value of χ^2, the greater is the discrepancy between the measured and assumed distributions. The test method is simple and it starts by calculating χ^2 using Eq. (3.22) and knowing the degrees of freedom v, table of Chi-square (χ^2) distribution or software are is consulted to find the probability $P(\chi^2)$. Nowadays, these tools are available on the internet. For the measured distribution to be accepted as a close (good fit) representation of the assumed (expected) distribution then $P(\chi^2)$ should fall in the range,

$$5\% \leq P\left(\chi^2\right) \leq 95\% \tag{3.23}$$

3.6 Errors and Uncertainty

Accuracy is a measure of the instrument's capability to truly specify the value of the quantity under measurement. Ideally, a measuring device will indicate the true value of the measured variable with 100% certainty. In reality, this is difficult to achieve, and an instrument will produce a value that deviates from the true (unknown) value within a range of uncertainty that is provided by the instrument's manufacturer. Many factors contribute to the uncertainty such as gain error, measuring noise, system dynamic noise, offset error, and instrument nonlinearity. These can be listed individually or summed into a single absolute accuracy specification. It is important to understand and account for all possible reasons of errors and uncertainties before accepting a measured value. However, accuracy cannot be improved beyond the resolution of the instrument.

Accuracy is not defined by a single type of deviation from the true value. There are several different sources of errors that contribute to the overall accuracy of the measuring device. *Error* is the deviation of the measurement from the true value of the quantity being measured given that the measurement is performed correctly. In other words, the accuracy of a measuring device is determined by considering a calculated margin of error as ±% of the full scale output (FSO).

Since it is usually not possible to know exactly what the true value is, an acceptable reference value for the quantity being measured must be defined following valid standards and procedures. A variety of factors may contribute to the deviation between multiple measurements of the same variable and between the average of the measurements and the value of the true or reference input. These factors may include variations in environmental conditions, time of measurement, calibration methods of transducer, measurement equipment, experimenters, or operators of equipment.

In general, repeated tests performed on presumable identical conditions do not yield identical measurement results. *Precision* is a measure of the closeness of the output readings to each other when produced from repeated measurements of the same input variable. It is usually expressed in terms of the standard deviation of the data population where the higher the standard deviation (variance) the less is the precision. There are two possible indicators of precision: repeatability and reproducibility. *Repeatability* refers to the instrument ability to produce the same output readings when repeating the experiment under extreme controlled conditions. Non-repeatability can be specified by determining the maximum deviation between the output readings, for the same input variables, as percentage of the full scale output. Therefore, repeatability is not associated with the true value of the measured variable but is rather a manifestation of the random scatter of the readings about their mean value. This is attributed to uncontrollable random errors resulting from unpredictable inherent factors in the measurement procedure that influence the output readings.

Reproducibility, on other hand, refers to the closeness of the output readings to each other when repeatedly measuring the same input variable under different testing conditions. Examples of such test conditions may include using different laboratories, instrumentation, or by different experimenters.

Signal noise or errors are combined of two types; random errors and systematic errors.

Systematic errors are consistent, fixed, and usually biased to one direction or the other with respect to the true value. Therefore, they primarily affect the accuracy of the measured variable by shifting the mean value of the measurements from the true value [4]. These errors are reproduceable by repetitive measurements. They are usually associated with incorrect calibration, biased equipment defects or design of experiment procedures or set up, and instruments wear. Incorrect use of equipment or overloading of sensors mechanically or electronically can lead to permanent defects or distortions that will cause consistent reading errors. Calibration adjustments and/or maintenance could be required to resolve these problems and restore a more accurate performance to the instrument. Some technical data sheets or catalogs are provided by manufacturers to direct the user on how to recalibrate or to repair the instrument. If the damage is beyond repair or recalibration then a replacement with a new instrument is required. Systematic errors are predictable and can be avoided or minimized by understanding how the measuring equipment works, by knowing the experiment constraints, and by following standards.

In contrast, **random errors** or variations are unpredictable and usually cause noise and inconsistencies in the measured results. These errors are usually unavoidable, have no pattern, and are irreproducible by repeat of experiment. The effects of random errors may be examined using statistical analysis methods. Unlike systematic errors, random errors primarily affect precision and cause the measured values to cluster around the sample mean value. Possible causes of random errors may include, environmental noise, instrument limitations and uncontrolled small variations in experimental procedures due to time of measurements or setups. Slight variations in sensor or target object mounting, position or alignment with respect to each other at each measurement instance may lead to this type of errors. For example, visual readings of gauges and sensors' scales, such as when taking several visual readings from different orientations. While systematic errors can be eliminated or minimized, random errors cannot be eliminated by adjustments to experiment procedure or devices. Random errors can be minimized by collecting more data points (increase sample size) and taking the average as an estimate of the true value.

Figure 3.9 illustrates the concepts of accuracy and precession and their relation to random and systematic errors. As depicted in this figure, random errors cause a scatter in the measured data about the mean. They may be caused by fluctuations in the sensor performance, instrumentation, the process under testing or its surrounding environment in ways that cannot be quantified using calibration or deterministic analytical techniques. Only through repeated data collection and statistical analyses that random errors can be estimated with some confidence interval (called random uncertainty). It is important to understand that both random and systematic errors affect the estimation of the accuracy.

Offset or zero shift error shown in Fig. 3.10 results when the instrument output shows a vertical intercept instead of reading a zero intercept when measuring a zero value for the input variable. Offset error results in a constant error for all measurements within the full scale. For example, a temperature transmitter for measuring 0–100 °C with an output of 4–20 mA, may have a zero-offset tolerance of ± 0.15 mA. In many applications, instruments are supplied with a zero-setting button or adjuster to nullify the offset error before start of measurements.

Sensitivity is usually defined as the ratio of the change in the sensor output to the corresponding change in the measurand value which is indicated by the slope of the calibration curve at a pint of interest. If the calibration curve is represented by a linear fit, then the sensitivity takes a constant value. The **sensitivity error** is a consistent change in the slope of the calibration curve. This may result in a proportionally skewed relation with increased or decreased slopes relative the best fit slope as shown in Fig. 3.11 for a linear calibration curve.

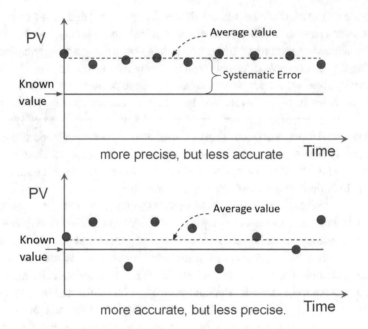

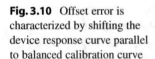

Fig. 3.9 Illustration of accuracy, precision, and systematic error

Fig. 3.10 Offset error is characterized by shifting the device response curve parallel to balanced calibration curve

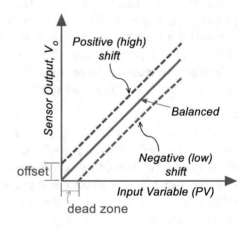

Non-linearity

Many real measuring systems do not exhibit a linear relationship between the input variable and the output reading. Devices calibrated using a linear curve fit as shown in Fig. 3.12 will indicate readings with a margin of error that can be specified by the linearity error [5]. This error which is a measure of the nonlinear behavior of the system is given by;

Fig. 3.11 Sensitivity error increases or decreases the slope of a linear calibration curve by a constant factor

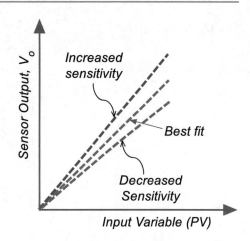

$$\%E_{\max\_Linearity} = \frac{\max[V_{o,real} - V_{o,Linear}]}{Full\ Scale\ (Output)} \tag{3.24}$$

Linearity error is defined as the maximum $\pm\%$ deviation of full-scale output (FSO) between the best fit straight line (BFSL) and the actual response curve (i.e., reference or characteristic curve). There are two commonly used methods of fitting a straight line to the data for linearity error estimation.

A line that is aligned midway between the two straight tangent lines that are enveloping all the measurements data of the calibration nonlinear curve is shown in Fig. 3.12 Another method establish a BFSL by determining the points on a the line that minimizes the sum of the squares of the deviations between the measurement output reading points and the corresponding points on the fitted straight line at the same variable input being measured. Calibration curve can be accepted with a zero or an offset 'intercept' at the starting point of the curve.

Fig. 3.12 A terminal best fit straight line and the linearity error

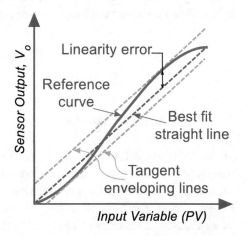

Fig. 3.13 Hysteresis error

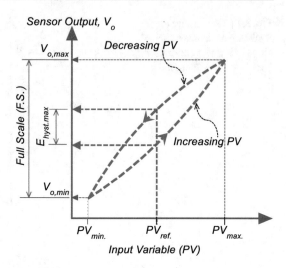

A straight line joining the extreme end data points of the transducer output curve is called the terminal linearity line. The linearity error in this type of fit is specified as the maximum deviation between the line and the curve as $\pm\%$ FSO. The end points are taken as the averaged end points data with the tolerance specified. Linearization can be achieved by using electronic circuits or software.

Hysteresis error (Fig. 3.13) refers to the difference in the measuring instrument output between increasing and decreasing change directions of the measured variable. It can be caused by the lag between the response of the sensing element and the corresponding change in the measured variable. Other causes include friction in sliding parts (e.g., potentiometer wiper) or time delays associated with capacitive or inductive electrical components within the measuring system. Hysteresis is defined as the maximum difference $\left(E_{hyst.\,\max}\right)$ in the output for any input value of the measured variable within the operating range of the device and is given, as a percentage of the full output scale, by

$$\%E_{hyst.\,\max} = \frac{E_{hyst.\,\max}}{Full\;Scale\;(F.S.)} \times 100 \tag{3.25}$$

3.6.1 Uncertainty Analysis

No matter how careful a competent (skilled) experimentalist experimenter is, both systematic and random errors (as discussed earlier in this chapter) will be present in all experiments. As the data contains different types of errors, it may be difficult to separate or isolate these errors. Furthermore, the components used in the design and fabrication of a measuring instrument will have differences that add up. Therefore, each instrument will have its own variations and uncertainty when taking measurements for the same PV. Performance of measurement systems should be expressed within validated margins of error. The discussion in this section aims at providing basic quantification methods for estimating the accuracy will include analyzing the data to quantify errors. It is important to validate the accuracy '±error range' that should be attached to any measured value or resulting quantities calculated using measured variables as inputs in a mathematical expression (function). Benefits of These types of analysis will lead to many such as error tracking, results validation, instrument selection, and design of experiment.

As discussed earlier in this chapter, the true value can best be estimated by the population mean. The population mean μ is related to the sample mean μ_s, given a probability level, by

$$\mu = \mu_s \pm [m \arg in \, of \, error evaluated data specified pd(x)\%]$$
$$= \mu_s \pm u_x$$

where u_x is the uncertainty associated with the variable x. Systematic and random errors must be determined for each variable x_i before proceeding with data analysis or calculating any results that depends on measured variables. These independent variables' uncertainties could be readily available in instrument's literature or data sheets, test codes, or by experience

3.6.2 Instrumentation Errors

The first type of error is related to the instrument resolution and is called the zero-order uncertainty, u_0. This error is independent of other sources of errors and is a random error caused by the scatter of data when taking readings while keeping all other measurement system parameters under control. This error is arbitrarily taken as half of the instrument resolution with probability of 95%. This means that 5% of the measured values (1 in 20) will may fall outside the interval defined by u_0. This error is related to digital quantization and is usually stated in the instrument catalog.

$$u_0 = \pm \frac{resolution}{2} \tag{3.26}$$

A second error type is associated with the instrument calibration u_c and is taken as the combination of systematic errors. These calibration characteristics are provided by manufacturers and are usually available in product catalogs or specifications sheets. The method of root-sum-squares (RSS) is practically used to determine the instrument uncertainty

$$u_c = \pm \sqrt{u_{c_1}^2 + u_{c_2}^2 + \cdots + u_{ck}^2} \tag{3.27}$$

where it is assumed that k sources of systematic errors are available for consideration. When performing these estimations, units and probability levels (e.g., 95%) should be kept consistent across all contributing factors in the analysis. These instrument calibration uncertainties can be used to compare between different experimental and instrumentation configurations so to choose the most accurate one [4, 6].

Other test operating conditions and control parameters of the measured variables are also considered as factors that can affect the accuracy of the measured value. Therefore, such paraments should be considered and are introduced as higher-order uncertainty estimates. These additional estimates give more conservative and more realistic estimate of the uncertainty in the results. As an example, the time of taking a measurement, setting all other test parameters at constant values, can influence the results due some hidden or minor factors that are time dependent in the process variable being measured. This requires collecting sufficient data and implementing a first-order statistical estimate of the uncertainty in the measured variable as a confidence interval given by

$$u_1 = \pm t_{v,pd}\sigma_s \tag{3.28}$$

The uncertainty margins are affected by the calibration precision and standards of the used instruments. Higher order errors could be associated with operating conditions such as sensor mounting, curve fitting methods, number of significant figures or resolution used in the numerical presentation of data. The Root sum squared (RSS) method combines all contributing errors to provide a more realistic estimate of the overall uncertainty as

$$u_{net} = \sqrt{u_0^2 + \sum_{i=1}^{k} u_{c_i}^2 + \sum_{j=1}^{M} u_j^2} \tag{3.29}$$

where,
$$u_j = \pm t_{j,v,pd}\sigma_{js}, \quad j = 1, 2, ..., M$$

3.6.3 Error Propagation

In many cases, measurement values of variables are substituted into mathematical functions to calculate the magnitude of some result of interest ($(\Re)$). A nominal or average value of $\Re$ can simply be calculated by using the average values of the measured variables. It is desirable in many applications to estimate and report the uncertainty in the calculated result $\Re$. The question is, what error range can be attached to the result? How do the errors in the measured variables contribute the $\pm$ error in the calculated result? Simple addition of the uncertainties assigned to the independent (measured) variables will yield an extreme combination that is highly not probable (unlikely) and can serve only as a preliminary rough estimate in some tests [7–10].

Let the result be given as a functional relationship of the independent measurements $x_1, x_2, ..., x_n$.

$$\Re = f(x_1, x_2, ..., x_n) \tag{3.30}$$

Kline and McClintock [11] showed that the uncertainty in the result $u_\Re$ for many engineering applications is given by the general form

$$u_\Re = \left[\left(\frac{\partial f}{\partial x_1} u_{x_1}\right)^2 + \left(\frac{\partial f}{\partial x_2} u_{x_2}\right)^2 + \cdots + \left(\frac{\partial f}{\partial x_n} u_{x_n}\right)^2\right]^{1/2} \tag{3.31}$$

For this relation to apply successfully, the variables x_i must follow a normal distribution and that the odds (i.e., probability levels) must be the same for each input variable uncertainty.

3.7 Regression Analysis

Calibration and experimental procedures involve taking several measurements of a dependent process variable (PV) while varying one or more independent variables under controlled conditions. Then regression analysis is used to establish the functional relationship between the measured (output, y_i) variables and the independent controlled (input, x_i) variables. In calibration procedures it is common to fix the value of the independent variable at some value x_i while taking several measurements of the dependent variable y, then the variable x is changed and fixed at another value and another set of measurements of y are taken, and so on. At each value x_i it is assumed that the y measurements follow a normal distribution. Selecting the optimum regression scheme and its parameters will lead to a successful and more accurate interpretation of the data. We can start this discussion by presenting the simple linear regression analysis of the form [8].

$$y' = ax + b \tag{3.32}$$

The method of least squares can be used to determine the parameters a and b that minimize the deviations between the measured data points y_i and the computed values y_i', where $i = 1, 2, ..., N$. The sum of the squares of the deviation is given by

$$E = \sum_{i=1}^{N} \left[y_i - y_i'\right]^2 = \sum_{i=1}^{N} \left[y_i - (ax_i + b)\right]^2 \tag{3.33}$$

Taking the derivatives of E with respect to a and b equal to zero gives

$$\frac{\partial E}{\partial a} = 0 \Rightarrow a \sum_{i=1}^{N} x_i^2 + b \sum_{i=1}^{N} x_i - \sum_{i=1}^{N} x_i y_i = 0$$

$$\frac{\partial E}{\partial b} = 0 \Rightarrow a \sum_{i=1}^{N} x_i + Nb - \sum_{i=1}^{N} y_i = 0 \tag{3.34}$$

Solving these two equations for the two unknowns a and b yields

$$a = \frac{N \sum_{i=1}^{N} x_i y_i - \sum_{i=1}^{N} x_i \sum_{i=1}^{N} y_i}{N \sum_{i=1}^{N} x_i^2 - \left(\sum_{i=1}^{N} x_i \right)^2} \tag{3.35}$$

and,

$$b = \frac{\sum_{i=1}^{N} y_i \sum_{i=1}^{N} x_i^2 - \sum_{i=1}^{N} x_i \sum_{i=1}^{N} x_i y_i}{N \sum_{i=1}^{N} x_i^2 - \left(\sum_{i=1}^{N} x_i \right)^2} \tag{3.36}$$

The errors between the measured points y_i and the calculated from regression function points y_i' can be presented by a single standard deviation estimate given by

$$\sigma_E = \sqrt{\frac{\sum_{i=1}^{N} \left[y_i - y_i' \right]^2}{v}} = \sqrt{\frac{\sum_{i=1}^{N} \left[y_i - (ax_i + b) \right]^2}{v}} \tag{3.37}$$

where the degrees of freedom of the linear regression fit $v = N - 2$ since the two parameters a and b are related. This procedure can be used to obtain higher order polynomial fits. A polynomial fit of order m for a single independent variable x is of the form

$$y' = b + a_1 x + a_2 x^2 + \cdots + a_m x^m \tag{3.38}$$

Again, the following sum of the squares equation is used to estimate the deviation between the measured and regression fitted values of the variable y

$$E = \sum_{i=1}^{N} \left[y_i - y_i' \right]^2$$

$$\Rightarrow E = \sum_{i=1}^{N} \left[y_i - \left(b + a_1 x_i + a_2 x_i^2 + \cdots + a_m x_i^m \right) \right]^2 \tag{3.39}$$

The $m + 1$ unknown coefficients $b, a_1, a_2, ..., a_m$ are determined by setting all partial derivatives of E with respect to these coefficients equal to zero as follows:

$$\frac{\partial E}{\partial b} = 0 \Rightarrow \sum_{i=1}^{N} (-2)[y_i - (b + a_1 x_i + a_2 x_i^2 + \cdots + a_m x_i^m)] = 0$$

$$\frac{\partial E}{\partial a_1} = 0 \Rightarrow \sum_{i=1}^{N} (-2x_i)[y_i - (b + a_1 x_i + a_2 x_i^2 + \cdots + a_m x_i^m)] = 0$$

$$\frac{\partial E}{\partial a_2} = 0 \Rightarrow \sum_{i=1}^{N} (-2x_i^2)[y_i - (b + a_1 x_i + a_2 x_i^2 + \cdots + a_m x_i^m)] = 0$$

.

.

.

$$\frac{\partial E}{\partial a_m} = 0 \Rightarrow \sum_{i=1}^{N} (-2x_i^m)[y_i - (b + a_1 x_i + a_2 x_i^2 + \cdots + a_m x_i^m)] = 0$$

The degrees of freedom for this mth order polynomial regression fit is $v = N - (m + 1)$ and the standard deviation error of the fit can be determined similar to Eq. (3.37).

$$\sigma_E = \sqrt{\frac{\sum_{i=1}^{N} [y_i - y_i']^2}{v}}$$

$$\Rightarrow \sigma_E = \sqrt{\frac{\sum_{i=1}^{N} \left[y_i - \left(b + a_1 x_i + a_2 x_i^2 + \cdots + a_m x_i^m \right) \right]^2}{v}} \tag{3.40}$$

3.8 Examples

Example 3.1 A small job shop produces motor shafts with a precision (standard deviation) of ± 0.5 mm for a large population. A quality control test is performed by randomly picking and measuring 10 shafts and the standard deviation of the tested sample is found to be 0.6 mm.

(a) Calculate the chi-square statistic for this test sample.
(b) Suppose the test is repeated with a new random sample of 10 motor shafts. What is the probability that the standard deviation in the new test would be greater than 0.6 mm?

Solution:

(a) Using Eq. (3.19a) the chi-square statistic is

$$\chi^2 = \frac{\nu\sigma_s^2}{\sigma^2} = \frac{(10-1)(0.6)^2}{(0.5)^2} = 12.96$$

(b) Knowing the degrees of freedom $N - 1 = 10 - 1 = 9$ and the chi-square statistic $\chi^2 = 12.96$, the cumulative probability can be obtained by consulting the chi-square table.

$$P(\chi^2 < 12.96) = 82\%$$
$$P(\chi^2 > 12.96) = 18\%$$

This means that the probability that the standard deviation for the new test would be greater than 0.6 mm is 0.18 or 18%.

Example 3.2 In a random sample of 10 aluminum plates, prepared for a heat treatment process, the mean thickness was 1 cm with a standard deviation of ± 0.1 mm. Determine the 5% significance level using the Student's t-distribution.

Solution:
Degrees of freedom $\nu = N - 1 = 10 - 1 = 9$.

Consulting the Student's t table with a 95% probability it is can be found that $t = 2.262$. Using Eq. (3.16)

$$\mu = \mu_s \pm \left(t_{\nu,p_d}\sigma_s\right) = \mu_s \pm \left(t_{\nu,p_d}\frac{\sigma}{\sqrt{N}}\right)$$

$$t_{\nu,p_d}\frac{\sigma}{\sqrt{N}} = 2.262\frac{(0.1)}{\sqrt{10}} = 0.0715$$

Thus, at a 95% confidence level the thickness of the aluminum plate $\approx 10 \pm 0.07$ mm.

Example 3.3 The following information is available from random pressure measurements in psi recorded using a digital pressure gauge connected to a data acquisition system. A hypothesis is made that the collected data follow a normal distribution with a confidence level of 95%. Test this hypothesis using the Chi-squared statistics.

Solution:

Group No.	Pressure range	Mean value, $\overline{x}_j$	$n_{j,m}$	$n_{j.exp}$	$\dfrac{(n_{j,m} - n_{j,\exp})^2}{n_{j,\exp}}$
1	68–69.9	68.9	5	4.58	0.0385
2	70–71.9	71.0	9	8.83	0.0033
3	72–73.9	72.8	13	8.19	2.825
4	74–76	75.0	6	5.42	0.1346
–	⇓ Given data			–	$\chi^2 = 3.00$

$$N = 5 + 9 + 13 + 6 = 33$$

$$\text{Sample mean} = \mu_s \frac{\sum\limits_{j=1}^{4} n_{j,m}\overline{x}_j}{N} = \frac{(5 \times 68.9 + 9 \times 71 + 13 \times 72.8 + 6 \times 75)}{33} = 72.1 \text{ psi}$$

The sample standard deviation is calculated using the following relation:

$$\sigma_s = \sqrt{\frac{1}{(N-1)} \sum_{j=1}^{4} n_{j,m}\overline{x}_j^2 - N\mu_s^2} = 2.5 \text{ psi}$$

The expected occurrence for the measurements in each group is calculated using Eq. (3.12) for the z_i limit and the Gaussian distribution.

For example, for group 1,

$$z_i = \frac{(x_i - \mu_s)}{\sigma_s} \Rightarrow \begin{cases} z_{68} = \dfrac{(68 - 72.1)}{2.5} = -1.64 \\ z_{69.9} = \dfrac{(69.9 - 72.1)}{2.5} = -0.88 \end{cases}$$

$$P(68 \le xi \le 69.9) = P(1.64) - P(0.88)$$
$$= 0.44950 - 0.31057 = 0.13893$$

$N = 33$, $n_{1,exp} = 33 \times 0.13893 = 4.58$

Other expected values for the remaining three groups are obtained similarly and are recorded in the table above.

$$\chi^2 = \sum_{j}^{4} \frac{(n_{j,measured} - n_{j,expected})^2}{n_{j,expected}} = 3.00$$

The degrees of freedom in the measured groups $v = M - k = 4 - 2 = 2$ where $k = 2$ is the number of constraints (the mean and standard deviation). Consulting a chi-squared table give a critical value of $\chi^2 = 5.99$ at a significant level of 5%. Therefore, since the observed $\chi^2 = 3.00$ is less than the critical $\chi^2 = 5.99$ there is no sufficient evidence to reject the hypothesis and the given data can be assumed to follow a normal distribution.

Example 3.4: A force sensor with a span of 0–2000 N measures a value of 950 N under some loading condition for a full scale output of 0–5 V. What is the sensor sensitivity? Determine the error and the actual measurement range for each of the following types of accuracy:

(a) ±0.65% of measuring span
(b) ±0.25% of FSO
(c) ±0.75% of reading

Solution:

$$\text{Sensitivity} = \frac{change\ in\ output\ signal}{change\ in\ measured\ variable} = \frac{5\ V - 0\ V}{2000\ N - 0\ N}$$
$$= 0.0025\ V/N,\ or\ 2.5\ mV/N$$

The error and actual measurement range,

(a) Span error = ±0.0065 × 2000 N = ±13 N. Thus the actual measurement will be in the range (950 − 13)–(950 + 13) N or 937–963 N.
(b) FSO error = ±0.0025 × 5 V = ±0.0125 V. This corresponds to a force error of ±0.0125 V × V/0.0025 V/N, or ±5 N. The actual measurement range is 945–955 N.
(c) Reading error = ±0.0075 × 950 N = ±7.125 N. The measurement is in the range 942.875–957.125 N or, 943–957 N.

Example 3.5 A digital pressure gauge is used to measure a nominal pressure of 75 psi. If the gauge specification sheet shows the following data at 95% confidence, provide an estimate of the pressure gauge uncertainty.

Pressure Range: 200 psi.
Sensitivity: 10 mV/psi.
Resolution: 0.5 psi.
Linearity error: 0.1% full range.
Hysteresis error: ±0.15% full range.
Accuracy: ±1.5% of reading.

Solution:
An estimate of the total uncertainty is determined by the root sum square (RSS) Eq. (3.29). First, the uncertainty of each contributing error is to be determined:
Resolution uncertainty (Eq. (3.26)),

$$u_0 = \pm \frac{resolution}{2} = \pm \frac{0.5}{2} = \pm 0.25 \, psi$$

Linearity uncertainty, $u_{c1} = (0.001 \times 200 \, psi) = 0.2 \, psi$.
Hysteresis uncertainty, $u_{c2} = (0.0015 \times 200 \, psi) = 0.3 \, psi$.
Accuracy uncertainty, $u_{c3} = (0.015 \times 75 \, psi) = 1.125 \, psi$.
An estimate of the overall uncertainty is given by Eq. (3.29) as

$$u_{net} = \sqrt{u_0^2 + \sum_{i=1}^{3} u_{c_i}^2}$$

$$u_{net} = \sqrt{(0.25)^2 + (0.2)^2 + (0.3)^2 + (1.125)^2}$$

$$u_{net} = \pm 1.21 \, psi$$

In terms of the indicated voltage, this is could be equivalent to $\pm (1.21 \, psi \times 10 \, mV/psi) = \pm 12.1 \, mV$ (95%). This analysis shows that the accuracy dominates the uncertainty in this measurement.

Example 3.6 The output of a measuring setup is modeled by $\mathfrak{R} = \frac{x_1^2}{x_2} + x_1 x_3$. Calculate the uncertainty in the resulting measurement given the following nominal values and uncertainties of the three variables x_1, x_2, and x_3:

$x_1 = 100 \, V, \, u_{x1} = \pm 1\%$
$x_2 = 1000 \, \Omega, \, u_{x2} = \pm 1.5\%$
$x_3 = 5 \, A, \, u_{x3} = \pm 0.5\%$

Solution:
The nominal result is

$$\mathfrak{R} = \frac{(100 \, V)^2}{1000 \, \Omega} + (100 \, V.5 \, A) = 510 \, W$$

To use Eq. (3.31) the derivatives are

$$\frac{\partial \mathfrak{R}}{\partial x_1} = \frac{2x_1}{x_2} + x_3 = \frac{(2 \times 100)}{1000} + 5 = 5.2$$

$$\frac{\partial \mathfrak{R}}{\partial x_2} = -\frac{x_1^2}{x_2^2} = -\frac{(100)^2}{(1000)^2} = -0.01$$

$$\frac{\partial \mathfrak{R}}{\partial x_3} = x_1 = 100$$

$$u_{x_1} = \pm 1.0 \, V, \, u_{x_2} = \pm 15 \, \Omega, \, u_{x_3} = \pm 0.025 \, A$$

$$u_\Re = \left[\left(\frac{\partial \Re}{\partial x_1}u_{x_1}\right)^2 + \left(\frac{\partial \Re}{\partial x_2}u_{x_2}\right)^2 + \left(\frac{\partial \Re}{\partial x_3}u_{x_3}\right)^2\right]^{1/2}$$

$$u_\Re = \left[(5.2 \times 1.0)^2 + (0.01 \times 15)^2 + (100 \times 0.025)^2\right]^{1/2}$$

$$u_\Re = [27.04 + 0.0225 + 6.25]^{1/2} = 5.77 \text{ W}$$

This is $(5.77/510) \times 100 = 1.13\%\Re$. The result can be stated as $\Re = 510 \pm 5.77$ W. It can be clearly seen that x_1 contributes the most to the result uncertainty, followed by x_3. The contribution from x_2 is the least significant.

3.9 Problems

3.1. A manufacturer used an ultrasonic precision thickness gauge to inspect the thickness of plastic packaging produced by a thermoforming process. Based on a large number of inspected samples, the variance in the thickness for one product is determined to be 0.005 mm. To meet quality requirements, the manufacturer selects and inspects 20 pieces at random and will reject any batch if the variance of the inspected pieces exceeds 0.007 mm. Determine the probability that a new batch with variance within tolerance might be rejected.

3.2. In a controlled process eight pressure measurements (in psi) are taken randomly within a period of 3 h and recorded as

85.1, 84.7, 84.9, 85.2, 84.9, 84.6, 85.3, and 84.8 psi.

Calculate the mean value, standard deviation, and tolerance limits for 90 and 95% confidence levels.

3.3. Given the following ten data points, use the method of least squares to determine a linear relationship $y = f(x)$.

i	1	2	3	4	5	6	7	8	9	10
x_i	1.0	2.5	3.3	4.2	5.0	5.9	6.4	7.7	8.6	9.5
y_i	0.85	1.8	3.0	4.5	5.3	5.1	7.0	8.1	9.3	10.4

3.4. During a temperature-controlled process 50 readings are recorded. The mean temperature is determined to be 75 °C and the standard deviation is 0.5 °C. How many readings are expected to fall within ±0.2 °C of the mean temperature with a confidence level of 95%? Solve this problem using the normal error distribution and the Student's t-distribution an compare results from both methods.

3.5. A testing laboratory ordered a large batch of 30 cm steel rods. It is stated by the suppliers that the standard deviation of the rod lengths is within ±2.0 mm. The laboratory technician inspected a random sample of 10 rods and found that the sample standard deviation is 2.5 mm. Determine the chi-square distribution and the associated level of significance.

3.6. A temperature transducer is used to measure a process temperature with nominal value of 150 °C. If the following data is available, calculate the total uncertainty of this temperature measuring set-up.

Transducer sensitivity $= 0.05$ mV/°C.

Full scale output voltage $= 0.5$ V.

ADC resolution error $= 0.12$ mV.

Reading uncertainty: $\pm 0.02\%$

Accuracy uncertainty: ± 1.0 °C.

Hysteresis error: $\pm 0.2\%$ FSO.

3.7. A pitot tube connected to a manometer is used to measure the pressure differential ΔP needed to calculate the fluid flow Q (m^3/s) in a tube. The following two equations are used

$$Q = A\sqrt{\frac{2\Delta P}{\rho_1}} \text{ and } \Delta P = \rho_2 g \Delta h$$

where A is cross section area of tube at the location of the pitot tube. Δh is the differential height of the measuring fluid in the manometer. A differential manometer is a device that measures the difference in pressure between two locations. If the following data is available, determine the nominal value of the flow rate and the uncertainty in the flow rate measurement.

$\Delta h = 20$ mm ± 1.0 mm

$\rho_1 =$ density of air $= 1.15 \pm 0.01$ kg/m^3

$\rho_2 =$ density of fluid in manometer $= 850 \pm 0.5$ kg/m^3

$g = 9.81$ m/s^2

$A = \frac{\pi}{4}D^2$ (where D is tube diameter).

$D = 5$ cm ± 1.0 mm.

References

1. Evans, M., N. Hastings, and B. Peacock. Statistical Distributions. 4th ed., John Wiley & Sons, Inc., Hoboken, NJ, 2011.
2. Everitt, Bian, and Skrondal, Anders, The Cambridge Dictionary of Statistics, 4th ed., Cambridge University Press, Cambridge, 2010.
3. Meeker, W. Q., and L. A. Escobar., Statistical Methods for Reliability Data., John Wiley & Sons, Inc., Hoboken, NJ, 1998.
4. Figliola, R.S., Beasley, D.E.: Theory ad Design for Mechanical Measurements, 7th ed.. Wiley, New York (2019)

5. Norton, Harry N., 'Sensor and analyzer handbook.' Prentice-Hall, Englewood Cliffs, N.J., (1982)
6. ASME PTC 19.1–1998, Test uncertainty, ASME. 1998.
7. H. W. Coleman and J. Steele, W. Glenn, Experimentation and Uncertainty Analysis for Engineers. Wiley, New York, 1999.
8. Holman, JP: Experimental Methods for Engineers, 5th ed. McGraw-Hill, New York (1989)
9. R. J. Moffat. Uncertainty analysis. In K. Azar, editor, Thermal Measurements in Electronic Cooling, pages 45–80. CRC Press, Boca Raton, FL, 1997.
10. R. J. Moffat. Using uncertainty analysis in the planning of an experiment. Journal of Fluids Engineering, 107:173–178, June 1985.
11. Kline, S. J., and McClintock, F. A., Describing Uncertainties in Single Sample Experiments, Mech. Eng., pp. 3–8, Jan. 1953.

Signal Conditioning Circuits and Devices

4

4.1 Introduction

To facilitate quality measurements, special electronic circuits, called signal conditioning circuits, are often required as interface between the transducers and the data acquisition systems. Examples of these circuits include input circuits, calibration circuits, operation amplifiers, and filters. It is possible that some of these signal conditioning circuits are integrated within the transducer, or they could be built in the data acquisition device.

Several sensors react to a change in the measurement parameter by changes in their own resistance R_s at the detector stage. At any measuring instant the sensor resistance of magnitude R_M can be measured using several special analog circuits. Three input circuits are presented in this section: current based circuit, voltage divider circuit, and the Wheatstone bridge.

4.2 Current-Based Input Circuit

A simple input circuit is shown in Fig. 4.1 in which a voltage source V_i with internal resistance R_i is connected to the changing sensor resistance. The Voltmeter V_M has a very large resistance and therefore is assumed to draw no current. Using ohms law and realizing that the two resistors R_i and R_M are connected in series, the current is determined by

$$i = \frac{V_i}{R_i + R_M} \quad \Rightarrow \quad \frac{i}{(V_i/R_i)} = \frac{1}{1 + (R_M/R_i)} \tag{4.1}$$

The voltage across the sensor measured by the Voltmeter is given by

© The Author(s), under exclusive license to Springer Nature Switzerland AG 2022
I. Abu-Mahfouz, *Instrumentation: Theory and Practice, Part 1*, Synthesis Lectures
on Mechanical Engineering, https://doi.org/10.1007/978-3-031-15246-7_4

Fig. 4.1 Current input circuit

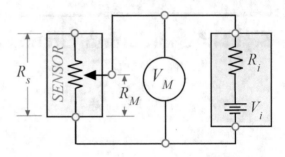

Fig. 4.2 Voltage and current characteristics for the current input circuit (Eqs. (4.1) and (4.2))

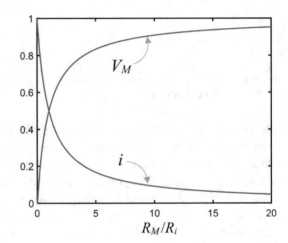

$$V_M = i\,R_M = \frac{V_i\,R_M}{R_i + R_M} \quad \Rightarrow \quad \frac{V_M}{V_i} = \frac{(R_M/R_i)}{1 + (R_M/R_i)} \tag{4.2}$$

The two nondimensional forms of Eqs. (4.1) and (4.2) are nonlinear as illustrated graphically in Fig. 4.2. However, it is more common to indicate the measurement from a sensor as a voltage signal. Therefore, the sensitivity of this circuit to changes in sensor resistance is given by

$$Circuit\ Sensitivity = \frac{\Delta V_M}{\Delta R_M} \approx \frac{dV_M}{dR_M} = \frac{V_i\,R_i}{(R_i + R_M)^2} \tag{4.3}$$

4.3   Voltage Divider Circuit

Figure 4.3 shows the basic voltage divider (potentiometer) circuit in which a constant and precise voltage V_i is applied across the transducer terminals. Assuming a very large internal resistance R_g of the voltmeter, the measured voltage corresponding to the variable sensor resistance R_M is

Fig. 4.3 Potentiometer (voltage divider) circuit

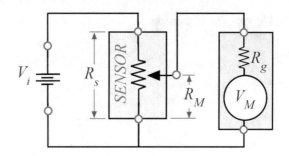

$$V_M = \left(\frac{R_M}{R_s}\right) V_i \tag{4.4}$$

Any current flow through the voltmeter affects the accuracy of the measured voltage V_M as an indicator of the measured variable. For a finite internal voltmeter resistance R_g, the source current in the potentiometer circuit is given by

$$i = \frac{V_i}{(R_s - R_M) + \left(\frac{R_g R_M}{R_g + R_M}\right)} \tag{4.5}$$

This loading condition will lead to the true voltage across RM being equal to

$$V_M' = i_M \cdot R_M = \left(\frac{R_g}{R_g + R_M}\right) \cdot i \cdot R_M$$

$$V_M' = \left(\frac{R_g R_M}{R_g + R_M}\right) \left(\frac{V_i}{(R_s - R_M) + \left(\frac{R_g R_M}{R_g + R_M}\right)}\right)$$

$$V_M' = \left(\frac{R_g R_M}{R_s R_M - R_M^2 + R_g R_s}\right) V_i$$

$$V_M' = \left(\frac{R_M}{R_s}\right) V_i \left(\frac{1}{\frac{R_M}{R_g} - \frac{R_M^2}{R_g R_s} + 1}\right)$$

$$V_M' = \left(\frac{R_M}{R_s}\right) V_i \left(\frac{1}{\frac{R_M}{R_g}\left(1 - \frac{R_M}{R_s}\right) + 1}\right) \tag{4.6}$$

4.4 Wheatstone Bridge

The Wheatstone bridge circuit presented in Fig. 4.4 provides accurate measurements of resistance changes of various types of transducers resulting from changes in the physical

Fig. 4.4 The Wheatstone bridge

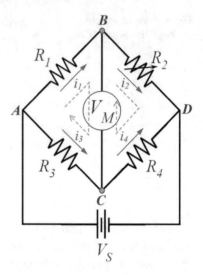

variable under measurement. The discussion in this section is limited to dc Wheatstone bridge. Wheatstone bridges for ac circuits are also available to measure capacitance and inductance changes for sensors that output these quantities in response to the physical variable under measurement. The change in resistance is translated as an output voltage by the bridge. Examples of such transducers are RTD, thermistors, and strain gages. One or more of the four resistors forming the bridge can be replaced by an equal number of the impedance of the equivalent circuit of equal number of transducers [1, 2]. The discussion here will be limited to the basic Wheatstone bridge as it presented in the schematic of Fig. 4.4. This bridge is commonly presented in this diamond shape for simplicity of analysis. In the following discussion R_1 is taken to represent a sensor's resistance.

4.4.1 Null Condition

For static, or slowly changing, measurements the Wheatstone bridge can be used in the Null mode. When the sensor resistance R_1 changes with the physical variable (PV), the Wheatstone bridge can be balanced ($V_{BC} = 0$) by adjusting a variable resistor such as R_2 in Fig. 4.4. The sensor R_1 can then be determined by using the null condition relation (Eq. (4.7)) and the physical variable value can be determined by knowing the sensitivity (transfer function or calibration chart) relationship of the sensor.

At Null condition, we can use Kirchhoff's voltage law for the loops ABC and BDC as

$$i_1 R_1 - i_3 R_3 = 0 \quad \text{and} \quad i_2 R_2 - i_4 R_4 = 0$$

and knowing that

$i_1 = i_2$ and $i_3 = i_4$ (No current through V_M)

The Null condition becomes:

$$R_1 R_4 = R_2 R_3 \qquad (4.7)$$

4.4.2 Deflection Mode

The deflection method is used for dynamic measurements of time varying physical variables. Making use of the voltage divider equation along the branch ABD and branch ACD of the bridge depicted in Fig. 4.4 we can write the following relations where V_s is the dc power supply voltage and V_M is the output measurement voltage,

$$V_B = Vs \left(\frac{R_1}{R_1 + R_2} \right)$$

$$V_C = Vs \left(\frac{R_3}{R_3 + R_4} \right)$$

$$V_M = V_B - V_C \Rightarrow V_M = Vs \left(\frac{R_1}{R_1 + R_2} - \frac{R_3}{R_3 + R_4} \right)$$

$$V_M = Vs \left(\frac{R_1 R_4 - R_2 R_3}{(R_1 + R_2)(R_3 + R_4)} \right) \qquad (4.8)$$

If we assume that a change in the physical variable will cause sensor R_1 to change its resistance by a magnitude ΔR_1. Then,

$$V_M = Vs \left(\frac{(R_1 + \Delta R_1)R_4 - R_2 R_3}{(R_1 + \Delta R_1 + R_2) \cdot (R_3 + R_4)} \right)$$

$$V_M = Vs \left(\frac{(R_1 + \Delta R_1)R_4 - R_2 R_3}{R_1(1 + \Delta R_1/R_1 + R_2/R_1) \cdot R_4(R_3/R_4 + 1)} \right)$$

$$V_M = Vs \left(\frac{1 + (\Delta R_1/R_1) - (R_2 R_3/R_1 R_4)}{[1 + (\Delta R_1/R_1) + (R_2/R_1)] [1 + (R_3/R_4)]} \right) \qquad (4.9)$$

It is a common practice to get a balanced Wheatstone bridge by choosing all four resistors to be equal to R and $V_M = 0$ initially (i.e. $R_1 = R_2 = R_3 = R_4 = R$). Substituting will give the following simplified but important characteristic relationship of the Wheatstone bridge can be obtained.

$$\frac{V_M}{Vs} = \frac{(\Delta R_1/R)}{[4 + 2(\Delta R_1/R)]} \qquad (4.10)$$

where the change in the sensor resistance R_1 is kept as ΔR_1. The output voltage of the Wheatstone bridge is typically expressed in millivolts output per volt input. As with other

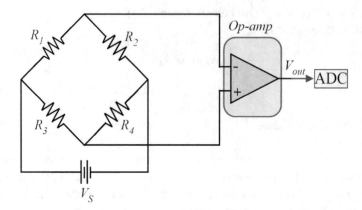

Fig. 4.5 An operational amplifier interface between the Wheatstone bridge and a digital to analog converter will improve signal quality

electronic component connections, to minimize the loading error (impedance mismatch), the resistance of the measurement instrument should be very large compared to the equivalent resistance of the bridge. Another method to ensure integrity of the measured potential between points B and C is to use an op-amp between the bridge output and the signal conditioning or converting devices such as an ADC or a DAQ as shown in Fig. 4.5.

4.5   Impedance Matching

When interfacing sensors to measuring devices, it is important to properly match the impedances (effective resistances) between the connected devices to minimize loading effects. Another important impedance matching consideration is between the load device and the power supply source to achieve max power utilization.

4.5.1   Signal Integrity

Figure 4.6 illustrates a voltage source V_s having output resistance R_s that is interfaced with a measuring device having a resistance R_L. The measured voltage V_L is the voltage drop across R_L and is given by:

$$V_L = \frac{R_L}{R_L + R_s} V_s \tag{4.11}$$

The measure value depends on the impedance ratio $\frac{R_L}{R_L + R_s}$, and therefore, the following measurement limits can be deduced,

Fig. 4.6 A simple source-load circuit to illustrate impedance matching

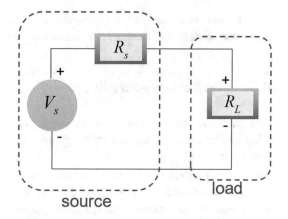

$$R_L << R_S \Rightarrow V_L \approx 0$$
$$R_L >> R_S \Rightarrow V_L \approx V_S$$

In general, it can be stated that for best signal integrity when interfacing two devices, the measuring device is recommended to have much larger input impedance than the source of the signal. Oscilloscopes, multimeters, and data acquisition devices have large impedance in the order of 1 MΩ or higher.

4.5.2 Maximum Power Consumption

The circuit shown in Fig. 4.6 and Eq. (4.11) can be used to discuss the case of optimizing power delivery. The power transmitted to the load from the supply source is given by,

$$P_L = \frac{V_L^2}{R_L} = \frac{R_L}{(R_L + R_s)^2} V_s^2 \tag{4.12a}$$

where, $V_L = \frac{R_L}{R_L + R_s} V_s$, is the voltage drop across the load. The relation between source and load impedances, R_L and R_s, for maximum power consumption can be obtained by setting the derivative of the power $P_L = 0$.

$$\frac{dP_L}{dR_L} = V_s^2 \frac{(R_L + R_s)^2 - 2R_L(R_L + R_s)}{(R_L + R_s)^4} = 0 \tag{4.12b}$$

Or,

$$(R_L + R_s)^2 = 2R_L(R_L + R_s) \Rightarrow R_L = R_s$$

Therefore, to maximize power transmission to a load, the load's impedance R_L should match the source's impedance R_s. Power supply sources have small impedance.

4.6   Operational Amplifiers

Operational amplifiers or op-amps are one of the most used signal conditioning circuits. They can be used to prepare output signals from sensors before they are input to other devices such as ADC and microcontrollers. They can also be used to manipulate output signals from microcontrollers so they can meet levels required by the controlled analog devices. This section will present the basic operation principles of the ideal op-amps and will discuss the commonly used practical op-amp circuits.

Figure 4.7 shows a schematic presentation of an ideal (open loop) op-amp with two inputs inverting (v_-) and noninverting (v_+) and a single output (V_{out}). Op-amps are active electronic devices and some literature omit the two power supply terminals $+ V_s$ and $- V_s$ from the schematic. The output voltage of an op-amp is limited to take values between these two power supply voltages with disregard to the values of the signal values at the input or the op-amp gain (K). V_{out} can be calculated as

$$V_{out} = K(v_+ - v_-) \tag{4.13}$$

The following three Ideal (open loop) op-amp assumptions are accepted.

1. The open loop op-amp gain is very large and is in the order of $K = 10^5$ or higher.
2. The op-amp draws no current due to the very large impedance at it input $(R_{i\infty} \geq 1\,\text{M}\,\Omega)$. This is very useful property for enabling accurate input voltage signals measurement.
3. The output impedance of an op-amp is very small (ideally zero). This assists in the signal integrity of the output voltage when connected to other components.

Fig. 4.7 Op-amp

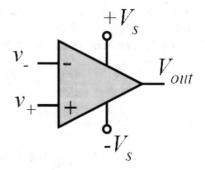

Fig. 4.8 Schematic of voltage follower op-amp

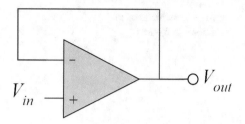

These three assumptions are invoked when analyzing op-amp circuits because they greatly simplify the analysis. However, real op-amps are seldom used in open-loop configuration and several more useful configurations of practical op-amps can be obtained by connecting other components such as resistors and capacitors to the op-amp. At least one fundamental connection is the feedback connection from the output of the op-amp to the inverting input (v_-) as shown in Fig. 4.8. In this case the op-amp operates in a closed-loop form with a gain that is defined by the components connected to the op-amp.

For an open-loop op-amp the voltage difference between the input voltages is virtually zero or it can be stated that the v_+ and v_- are equal virtually. This virtual rule can be deduced rom equation == knowing that the output voltage V_{out} will not exceed the limits of the supplied voltage. With very large gain, the difference between the two input voltages $(v_+ - v_-) = (V_s/100{,}000) \approx 0$. This rule also applies to other derived real op-amps. In the following we will discuss the operating characteristics of some commonly used practical op-amps [3–5]. In the analysis we will make use of the assumptions discussed above.

4.6.1 The Buffer OP-AMP

The voltage following op-amp is a unity gain buffer op-amp that in which the input signal is connected to the noninverting terminal and the output (V_{out}) is connected to the inverting terminal via a zero-resistance feedback ($V_{out} = v_-$). This configuration will lead to an-amp with a gain of $K = 1$ (Fig. 4.8) since $v_- = v_+$ and $V_{out} = v_-$, then $V_{out} = v_+ = V_{in}$. This op-amp is very desirable where impedance shielding is required between components in circuits to preserve the stabilize the voltage level of the source signal and to protect against signal distortions that may be caused by impedance mismatch.

4.6.2 The Non-inverting OP-AMP

The basic circuit for the noninverting op-amp is shown in Fig. 4.9. To find the gain equation of this op-amp can apply Kirchhoff current law at the voltage divider junction between R_f and R_g in addition to invoking the simplifying op-amp rules discussed earlier

Fig. 4.9 Schematic of the noninverting op-amp

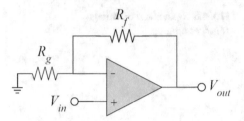

in this section. The first rule states that no current is drawn into the op-amp. This will lead to $i_f = i_g$ in the feedback loop. The second rule states that $v_- = v_+$ and therefore $v_- = V_{in}$ virtually. Using Ohm's law the rest of the analysis is as follows:

$$i_g = \frac{v_- - 0}{R_g} = \frac{V_{in} - 0}{R_g}$$

and

$$i_f = \frac{V_{out} - v_-}{R_f} = \frac{V_{out} - V_{in}}{R_f}$$

$$i_g = i_f \Rightarrow \frac{V_{in} - 0}{R_g} = \frac{V_{out} - V_{in}}{R_f}$$

$$\text{Solving for } V_{out}, \Rightarrow V_{out} = \left(1 + \frac{R_f}{R_g}\right) V_{in} \tag{4.14}$$

It can be clearly seen from this relation that the gain for this op-amp is positive (noninverting) and is greater than unity.

4.6.3 The Inverting OP-AMP

By allowing the input signal to be connected to R_g (now R_i) and by connecting the noninverting terminal to ground a different op-amp is created as shown in Fig. 4.10. For convenience we have changed the name of R_g to R_i where the input signal V_{in} is connected. Following a similar analysis procedure as for the noninverting op-amp the following relations apply to the inverting op-amp.

$$v_- = v_+ = 0,$$

$$i_i = \frac{V_{in} - v_-}{R_i} = \frac{V_{in} - 0}{R_i}$$

and

Fig. 4.10 Schematic of the inverting op-amp

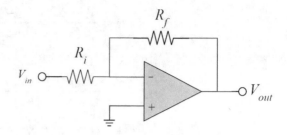

$$i_f = \frac{v_- - V_{out}}{R_f} = \frac{0 - V_{out}}{R_f}$$

$$i_i = i_f \Rightarrow \frac{V_{in} - 0}{R_i} = \frac{0 - V_{out}}{R_f}$$

$$Solving\ for\ V_{out},\ \Rightarrow V_{out} = -\left(\frac{R_f}{R_i}\right)V_{in} \tag{4.15}$$

The inverting op-amp reverses (inverts) the polarity of the input signal and has a gain that is defined as the ratio between R_f and R_i ($K = -R_f / R_i$).

4.6.4   The Summing OP-AMP

The summing (summer) op-amp, as the name indicates, produces a single output voltage that is the sum of the input voltages. Figure 4.11 shows the circuit for the summer op-amp with only two inputs. The results of the following analysis can be generalized to include op-amps with more than two inputs. Again, we follow similar procedure as before by invoking the simplifying rules of the op-amp.

$$v_- = v_+ = 0,$$

$$i_1 = \frac{V_1 - v_-}{R_1} = \frac{V_1 - 0}{R_1} = \frac{V_1}{R_1}$$

Fig. 4.11 Schematic of the summing op-amp

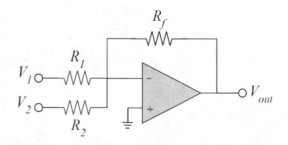

$$i_2 = \frac{V_2 - v_-}{R_2} = \frac{V_2 - 0}{R_2} = \frac{V_2}{R_2}$$

and

$$i_f = \frac{v_- - V_{out}}{R_f} = \frac{0 - V_{out}}{R_f} = -\left(\frac{V_{out}}{R_f}\right)$$

$$i_1 + i_2 = i_f \Rightarrow \frac{V_1}{R_1} + \frac{V_2}{R_2} = -\left(\frac{V_{out}}{R_f}\right)$$

Solving for V_{out}:

$$V_{out} = -R_f\left(\frac{V_1}{R_1} + \frac{V_2}{R_2}\right) = -\left(\frac{R_f}{R_1}\right)V_1 - \left(\frac{R_f}{R_2}\right)V_2 \qquad (4.16)$$

If, as a special case, we take

$R_1 = R_2 = R_i$, then the expression for V_{out} can be written as

$$V_{out} = -\frac{R_f}{Ri}(V_1 + V_2) \qquad (4.17)$$

Hence, Equation (4.16) shows that each ith input voltage signal V_ith can have its own gain value defined by the ratio between the feedback resistor R_f and its respective input resistor R_ith. It is clear that the summing op-amp is a kind of an inverting op-amp with more than one input.

4.6.5 The Differential Op-Amp

The differential (difference) op-amp (Fig. 4.12) produces a single output voltage V_{out} that is proportional to the difference between the two voltage signals connected at it input terminals. Let's follow the following relations to derive the gain equation (transfer function) for this important op-amp!

Using the voltage divider relation to get the voltage at V_+

$$v_+ = V_2\left(\frac{R_g}{R_2 + R_g}\right) = v_- \quad (because, \ v_- = v_+),$$

$$i_1 = \frac{V_1 - v_-}{R_1} = \frac{V_1 - V_2\left(\frac{R_g}{R_2 + R_g}\right)}{R_1}$$

and

Solving for V_{out}:

Fig. 4.12 The differential
op-amp

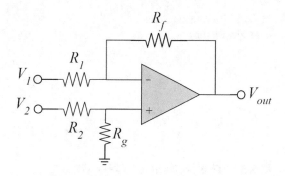

$$V_{out} = -R_f \left[\left(\frac{V_1}{R_1} \right) - V_2 \left(\frac{R_1 + R_f}{R_1 R_f} \right) \left(\frac{R_g}{R_2 + R_g} \right) \right] \qquad (4.18)$$

order for the transfer function to actually calculate the difference between the two input signals, we must assume $R_1 = R_2 = R_i$ and $R_g = R_f$, then the expression for Vout can be
Written as

$$V_{out} = \frac{R_f}{Ri} (V_2 - V_1) \qquad (4.19)$$

It is worth noting here that the input signals to the differential op-amp are connected to input resistors that may not be high enough, compared to the impedance of the signal source component, and therefore the signal may be degraded (Sect. 5.1). To overcome this problem, buffer op-amps are placed between the input signals and the differential op-amp as shown in Fig. 4.13. This configuration is called 'Instrumentation amplifier' and it is very commonly used in test and measurement practices. The output of the difference op-amp (or that of the instrumentation amplifier) is proportional to only the difference between the two input voltage signals with disregard to their respective values relative to ground.

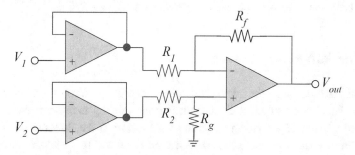

Fig. 4.13 Schematic of the instrumentation op-amp

Fig. 4.14 Schematic of the
derivative op-amp

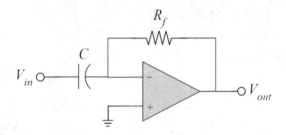

4.6.6   The Derivative OP-AMP

This op-amp produces an output that is proportional to the rate of change of the input signal with respect to time. Therefore, this op-amp is not to be confused with the differential
op-amp which amplify the difference between two input signals. Figure 4.14 shows the
basic circuit for the derivative op-amp used in the following analysis. Applying Kirchoves
current law to the node at the inverting terminal we get $i_c = i_f$.

$v = v_+$ (*virtual ground*)

The current for the capacitor branch can be determine by,

$$i_c = C \cdot \frac{dV_c}{dt}$$

and the current through the feedback resistor R_f is

$$i_f = \frac{v_- - V_{out}}{R_f} = \frac{0 - V_{out}}{R_f}$$

$$i_c = i_f \Rightarrow C \cdot \frac{dV_c}{dt} = \frac{0 - V_{out}}{R_f}$$

$$V_{out} = -Rf_c \cdot \frac{dV_c}{dt} \tag{4.20}$$

As can beby seen this op-amp has a negative gain ($K = -R_f C$).

4.6.7   The Integrator Op-Amp

There are applications were the integral quantity over a period time of a measured variable
is required. Example is to get a displacement from velocity measurements or to get an
accumulated amount of a flowing material over time from the flow rate measurements,
etc. When the signal of time dependent measurement is fed to an integrator op-amp the
output will be the time integral of the measurement. This is equal to the area under the
signal-time curve as shown in Fig. 4.15. Again, applying Kirchoves current law to the
node at the inverting terminal we get $i_i = i_c$.

Fig. 4.15 The integrator op-amp

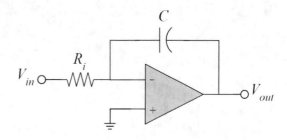

$$v_- = v_+ = 0 \quad (virtual\ ground)$$

The current for the capacitor branch can be determine by,

$$i_c = C \cdot \frac{dV_c}{dt},$$

and the current through the input resistor R_i is

$$i_i = \frac{V_{in} - v_-}{R_i} = \frac{V_{in} - 0}{R_i}$$

$$i_c = i_i \Rightarrow C \cdot \frac{dV_c}{dt} = \frac{V_{in}}{R_i}$$

where

$$V_c = v_- - V_{out} = 0 - V_{out} \Rightarrow V_c = -V_{out}$$

$$C \cdot \frac{-dV_{out}}{dt} = \frac{V_{in}}{R_i} \Rightarrow \int_{V_{out}(0)}^{V_{out}} dV_{out} = -\frac{1}{R_i C} \int_0^t V_{in} \cdot dt$$

Hence,

$$V_{out}(t) = -\frac{1}{R_i C} \int_0^t V_{in} \cdot dt + V_{out}(0) \tag{4.21}$$

where $V_{out}(0)$ is the initial output value at start of integration. Integrator Op-amps needs frequent resetting because their output reach saturation if integration results reach its operational limits.

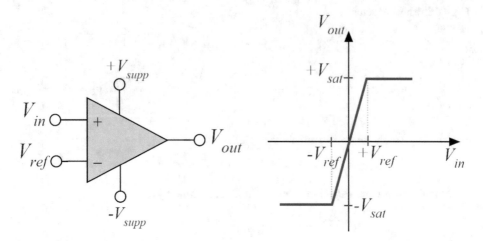

Fig. 4.16 The comparator op-amp

4.6.8 The Comparator OP-AMP

Due to its very large gain the output voltage of the comparator open-loop op-amp will saturate at the one of the two supplied voltage limits following the input terminal sign with the more positive input. Therefore, this op-amp (Fig. 4.16) operates as an automatic electrical switch based on its input signals. As a numerical illustration of this concept, if the input to the noninverting terminal is $+ 3.7$ V and the input to the inverting terminal is $+ 4$ V then using Eq. (4.13) $V_{out} = K (3.7 - 4) = K (- 0.3)$, and with K very high $\rightarrow V_{out} = -V_{sat}$.

4.6.9 Voltage-To-Current Converter

The circuit in Fig. 4.17 utilizes a non-investing op-amp to convert a voltage signal (form a sensor) into current that can be transmitted over long distances without degradation. Industrial standard is the 4–20 mA current signal that is linearly proportional to the original voltage range. A simple resistor at the receiving end is used to convert the current back to voltage signal. The current i is independent of the load resistance R_L.

$$i = \frac{V_{in}}{R_g} \quad \Rightarrow \quad V_{out} = i(R_L + R_g) \tag{4.22}$$

Fig. 4.17 Voltage-to-current converter circuit

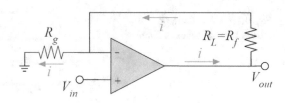

Fig. 4.18 Current-to-voltage converter circuit

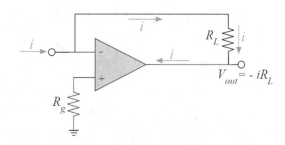

4.6.10   Current-To-Voltage Converter

Some sensors produce current signals representing the measured variables so they can be transmitted over long lines without losing their current information. At the receiving point, these signals may need to be converted to voltage values proportional to the original measured variable. Figure 4.18 shows a simple op-amp circuit that converts a current signal i into an output voltage V_{out} across the load resistance R_L.

4.7   Filters

Measurement signals of physical variables usually contain more information than that directly related to the quantity under measurement. These additional components of information are considered noise as they are not of significance to the sought-after data set. An experimenter taking measurements of the air inside a room may find high frequency fluctuations in the data. These noisy contributions may come from instrumentation or environmental sources. A priori knowledge about the frequency range of the desired variable will aid in identifying the magnitude and dynamics of the noise component present in the acquired data.

Suppressing or filtering out noisy information can be achieved using filter circuits. This section presents some basic analog filter circuits that are commonly used to improve signal quality. Digital filtering using software is also effective in cleaning data and eliminating noise. More on filters, their circuits design and characteristics can be found in many specialized references on filters (see for example [6]).

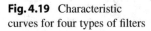

Fig. 4.19 Characteristic
curves for four types of filters

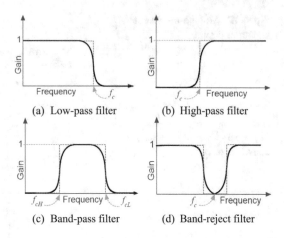

(a) Low-pass filter   (b) High-pass filter

(c) Band-pass filter   (d) Band-reject filter

Figure 4.19 shows a close-to-ideal characteristics for four types of filters with Gain = $|V_{out}/V_{in}|$. Practical filters do not have a crisp (sharp) transition between the passband and stopband. There is a range of frequency over which the filter gain changes gradually with frequency. A low-pass filter passes signals with low frequency and suppresses those with high frequency. A cutoff frequency (f_c) is a critical value that separates the two bands of frequencies. For a simple RC filter f_c is defined by

$$f_c = \frac{1}{2\pi RC} \tag{4.23}$$

A high-pass filter does the opposite, it permits components in the signal with frequencies higher than f_c and blocks components with frequencies lower than f_c. The critical frequency is also defined by Eq. (4.23). The gain magnitude at the cutoff frequency is approximately 0.707. This corresponds to an output amplitude attenuation by 3 dB from the input signal amplitude. To improve the transition sharpness around fc, two or more similar filters can be cascaded in series. The net gain of the set of n cascaded filters is equal to the single filter gain raised to the power n.

Two circuits for the passive low-pass and high-pass RC filters are shown in Fig. 4.20.

A band-pass filter permits components with frequencies within a band defined by two limits f_{cH} and f_{cL} and rejects all those with frequencies higher or lower than the permitted band. This can be achieved by a low-pass and a high-pass filters cascaded in series with the condition that $f_{cH} < f_{cL}$. The filter with the lower f_{cH} is a high-pass filter and the one with higher f_{cL} is a low-pass filter.

Opposite of a band-pass filter is a band-reject (or notch) filter which rejects a narrow band of frequencies and permits all other frequencies higher and lower than the rejected band. Notch filters are often used to reject a single noise frequency in the measured signal.

Passive filters characteristics and gains can be enhanced when integrated with operations amplifiers. Active filters are combinations of passive filters with op-amp circuits

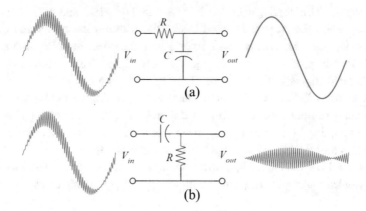

Fig. 4.20 Circuit for the **a** low-pass and **b** high-pass filters with illustration of input and output signals

Fig. 4.21 Basic circuit for a non-inverting low-pass active filter

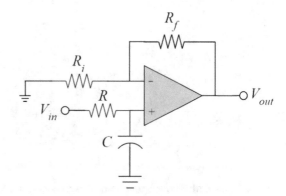

to achieve desired performance. A low-pass active is shown in Fig. 4.21. Using a non-investing op-amp. This filter has a gain similar to that of the inverting op-amp ($K = R_f / R_i$) and the cutoff frequency as defined by Eq. 4.23.

4.8   Schmitt Trigger

Because of their very high gain and fast response, comparators are sensitive to input signal noise which will be magnified causing undesirable transitions in the response of the logic connected at their output. The effect of noise in the input signals can be minimized and the output can be stabilized by using a Schmitt trigger. A Schmitt trigger is a type of a comparator with a resistor network that is configured with positive feedback that shifts the single switching threshold level of a comparator into two threshold levels. This causes a favorable delay in the output response relative to noisy changes in the input signal. A simple inverting Schmitt trigger circuit is shown in Figs. 4.22 and 4.23 illustrates one example of its delayed response (called hysteresis) between increasing and decreasing paths of the input signal. This is useful in removing the effect of noise from the analog input signal while producing a stable digital output signal. The operation of the non-inverting Schmitt trigger is similar in effect but opposite in polarity.

For the inverting Schmitt trigger circuit shown in Fig. 4.22, the feedback loop through R_3 is between the output voltage and the input to the non-inverting terminal of the op-amp. With the assumption of a small $R_{pull\text{-}up}$ resistor relative to R_3, the analysis is simplified by separately considering the two output levels of high and low.

When $V_{in} > V_{ref}$, the output V_{out} will saturate low (or drop to 0 V) and R_3 right-end will be grounded making R_3 in parallel with R_2 ($R_3 \parallel R_2$) as depicted by the left circuit

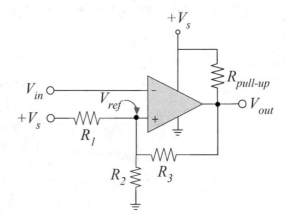

Fig. 4.22 Inverting comparator with hysteresis as an inverting Schmitt trigger

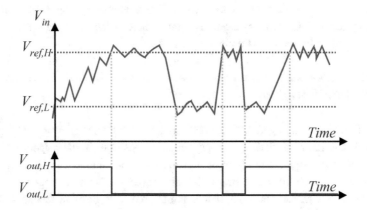

Fig. 4.23 Example response plot for inverting Schmitt trigger

in Fig. 4.24. This will cause lower threshold voltage $V_{ref,L}$ to be given by,

$$V_{out} = 0 \quad (low)$$

$$V_{ref,L} = \frac{R_2 \| R_3}{R_1 + (R_2 \| R_3)} V_s \qquad (4.24)$$

V_{in} needs to decline below the lower threshold $V_{ref,L}$ for V_{out} to switch to $V_{out,H}$.

On the other hand, when $V_{in} < V_{ref}$, the output V_{out} will saturate high (or $+V_s$) and R_3 right-end will be at $+V_s$ leading to $R_3 \| R_1$, as depicted by the circuit on the right in Fig. 4.24. This will cause the upper threshold voltage $V_{ref,H}$ to be given by

$$V_{out} = V_s \quad (high)$$

$$V_{ref,H} = \frac{R_2}{R_2 + (R_1 \| R_3)} V_s \qquad (4.25)$$

V_{in} needs to rise above the upper threshold $V_{ref,H}$ for V_{out} to switch to 0 V or $V_{out,L}$.

As illustrated in Fig. 4.23, when the noise level in the input signal is less than the difference between the two hysteresis thresholds ($V_{ref,H} - V_{ref,L}$), the output of the op-amp will not change until the input rises $V_{ref,H}$ or falls below $V_{ref,L}$. The two thresholds in the Schmitt trigger will dissipate the noise influence on the switching action of this op-amp. Schmitt triggers are widely used in logic circuits and as input interfaces to microprocessors to eliminate undesirable transitions caused by noisy changes in the input signals. Schmitt triggers are also used in switch debouncing which is the elimination of the jitter (bouncing) produced by actuating mechanical switches. Figure 4.25 shows a generalized plot of the transfer function with hysteresis for the inverting Schmitt trigger circuit.

Fig. 4.24 Symbol of the hysteresis response in the inverting Schmitt trigger

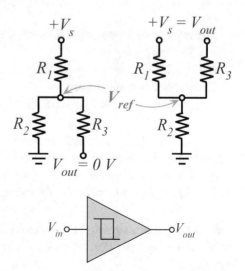

Fig. 4.25 Transfer
Characteristics of an inverting
Schmitt trigger

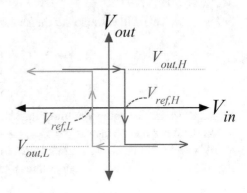

4.9   Examples

Example 4.1 Design a differential amplifier that accepts a signal from a displacement sensor
at the non-inverting terminal. The op-amp output should range from 0 V to +2.5 V for a
change in the sensor signal in the range from 2 to 7 V. The sensor has an effective resistance
of 1.5 kΩ.

Solution
Assuming linear characteristics ($V_{out} = KV_{in} + b$) for the op-amp, the relation between
input and output is determined by the two (*input, output*) limit points (2, 0) and (7, 2.5 V).
The gain of the op-amp K is calculated by

$$K = \frac{2.5\ \text{V} - 0\ \text{V}}{7\ \text{V} - 2\ \text{V}} = 0.5$$

The value of the intercept b is determined to satisfy the linear relation at any of the limit
points. The first point gives
  0 V = 0.5(2 V) + b → b = −1 V.
  Therefore, the op-amp relation is
  $V_{out} = 0.5V_{in} - 1$ V, or $V_{out} = 0.5(V_{in} - 2)$.

This op-amp (Fig. 4.12) needs a constant 2 V (set-point) input to the inverting terminal
which can be provided by any 2 V source or by using a voltage divider to obtain the
required 2 V from other DC sources. The following resistors can be used to have a gain
of $K = R_f / R_1 = 0.5$ by selecting R_f and $R_g = 100$ kΩ and $R_1 = R_2 = 50$ kΩ.

Example 4.2 Calculate and plot the output V_{out} of an integrator op-amp (shown in Fig. 4.14)
for the following input signal. Use $R_{in} = 100$ kΩ, and $C = 10\ \mu F$ (Figs. 4.26 and 4.27).

Fig. 4.26 Input signal for
Example 4.2

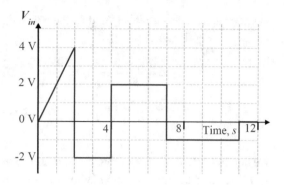

Fig. 4.27 Output solution for
example 4.2

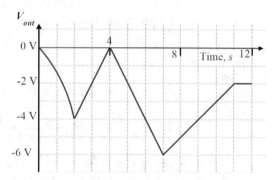

Solution

Using Eq. (4.13), the gain $(-1/RC) = -1$, and V_{out} can be calculated at key instants in time, where V_{in} changes its form, as follows.

$$V_{out}(2s) = 0 + [(-1) \cdot (0.5)(2 \times 4)] = -4 \text{ V}.$$
$$V_{out}(4s) = -4 + [(-1) \cdot (2 \times -2)] = 0 \text{ V}.$$
$$V_{out}(7s) = 0 + [(-1) \cdot (3 \times 2)] = -6 \text{ V}.$$
$$V_{out}(11s) = -6 + [(-1) \cdot (4 \times -1)] = -2 \text{ V}.$$
$$V_{out}(12s) = -2 + 0 = -2 \text{ V}.$$

Example 4.3 In light of the discussion for the inverting Schmitt trigger circuit shown in Fig. 4.22, discuss the operation of the following non-inverting Schmitt trigger and plot its output signal V_{out} for the input signal shown in Fig. 4.23.

Solution

This is a non-inverting comparator op-amp with positive feedback through R_2 and an output response with hysteresis. The voltage level at the non-inverting terminal junction is virtual ground ($V = 0$ V), and the relation between V_{in} and V_{out} can be obtained by using ohms law to calculate the currents i_1 and i_2. Note that since no current is drawn by the op-amp circuit, $i_1 = i_2$ (Fig. 4.28).

$$i_1 = \frac{V_{in} - 0}{R_1} = i_2 = \frac{0 - V_{out}}{R_2} \quad V_{ref,H} = -\frac{R_1}{R_2} V_{out,L}$$

$$\Rightarrow \frac{V_{in}}{R_1} = \frac{-V_{out}}{R_2} \quad \Rightarrow \quad V_{ref,L} = -\frac{R_1}{R_2} V_{out,H}$$

Example 4.4 In reference to Fig. 4.6, a sensor with an output voltage of 5 V and internal resistance of 5 kΩ is connected directly to an inverting op-amp (shown in Fig. 4.10) with internal resistance of $R_i = 10$ kΩ. Determine the voltage drop across the amplifier's internal resistance R_i and suggest a solution to reduce the loading effect in this circuit (Figs. 4.29 and 4.30).

Solution
Using Eq. (4.11) with $R_L = R_i = 10$ kΩ and $R_s = 5$ kΩ, the voltage drop V_L across R_i is,

$$V_L = \frac{R_L}{R_L + R_s} \ V_s = \frac{10 \text{ k}\Omega}{10 \text{ k}\Omega + 5 \text{ k}\Omega} 5 \text{ V} = 3.3\overline{3} \text{ V}$$

Hence, due to loading effects the received sensor signal at the op-amp is 3.33 V and not the original 5 V, a deviation of 33.3%. This loading effect can be minimized by using an op-amp with higher input impedance R_i. For example, if $R_i = 100$ kΩ, the same calculation gives a $V_L = 4.76$ V (a 4.8% deviation). The higher the op-amp input impedance the less the error due to loading effects. Inserting a voltage follower between the sensor output and the op-amp input could be a better solution without the need to have an op-amp with higher R_i. The voltage follower draws no current from the sensor output signal and the 5 V will appear at the voltage follower output. The 10 kΩ input resistance of the inverting op-amp is much higher than the output resistance of the voltage follower and no loading effect will take place there too.

Fig. 4.28 Non-inverting comparator with hysteresis as a non-inverting Schmitt trigger (Example 4.3)

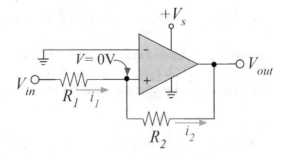

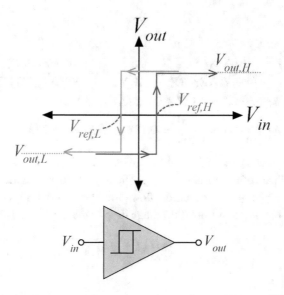

Fig. 4.29 Non-inverting Schmitt trigger characteristic function and symbol (Example 4.3)

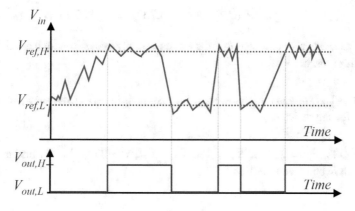

Fig. 4.30 Non-inverting Schmitt trigger response plot (Example 4.3)

Example 4.5 For the Wheatstone bridge shown in Fig. 4.4, initially all four resistors are equal to 350 Ω and the bridge is balanced. A supply voltage of $V_s = 12$ V is connected between nodes A and D. If R_1 is a sensor (such as a strain gauge) that changes with a measured physical variable (such as strain), determine the change in R_1 (ΔR_1) if the output voltage reading is $V_M = 0.5$ V. Also, determine the amount of current flow and the power dissipation through the sensor at this instant.

Solution

Using Eq. (4.10)

$$\frac{V_M}{Vs} = \frac{(\Delta R_1/R)}{[4 + 2(\Delta R_1/R)]} \Rightarrow \frac{0.5 \text{ V}}{12 \text{ V}} = \frac{(\Delta R_1/350 \text{ } \Omega)}{[4 + 2(\Delta R_1/350 \text{ } \Omega)]}$$

$$(\Delta R_1/350 \text{ } \Omega) = 0.167 + 0.083(\Delta R_1/350 \text{ } \Omega)$$

$$(\Delta R_1/350 \text{ } \Omega) = \frac{0.167}{0.917} = 0.182 \Rightarrow \Delta R_1 = 63.7 \text{ } \Omega$$

and the total sensor resistance is $R_1 + \Delta R_1 = 413.7 \text{ } \Omega$.

For the second part of this example, a high impedance measuring device is assumed to be used ($R_M \approx \infty$) then $i_M = 0$ (no current flow through the measuring device). The current i_1 flowing through R_1 is given from the voltage drop in the upper arm (Fig. 4.4).

$$V_s = i_1 R_1 + i_2 R_2.$$

And since $i_M = 0, i_1 = i_2$.

$i_s = Vs/(R_1+R_2) = 12$ V/(413.7 Ω+350 Ω) =0.0157 A.or, $i_1 = 16$ mA.

Power dissipation

$$P_1 = i_1^2 R_1 = (0.016 \text{ A})^2 (413.7 \text{ } \Omega) \Rightarrow P_1 = 0.11 \text{ W}$$

In this case, it is also assumed that the voltage supply is stable with negligible internal battery voltage losses.

Example 4.6 The voltage impressed on a Wheatstone bridge (Fig. 4.4) is $V_s = 24$ V, and it uses the following resistors:

$R_1 = 300 \text{ } \Omega, R_2 = 200 \text{ } \Omega, R_3 = 400 \text{ } \Omega$, and $R_4 = 400 \text{ } \Omega$. Is this bridge balanced? Determine the output voltage V_M.

Solution

For this Wheatstone bridge to be balanced the following relation (Eq. (4.7)) must be satisfied.

$R_1 R_4 = R_2 R_3 \rightarrow (300 \times 400) \neq (200 \times 400)$.

$120,000 \neq 80,000 \rightarrow$ Bridge not balanced.

The output voltage can be determined using Eq. (4.8)

$$V_M = Vs \left(\frac{R_1}{R_1 + R_2} - \frac{R_3}{R_3 + R_4} \right)$$

$$V_M = 24 \text{ V} \left(\frac{300 \text{ } \Omega}{300 \text{ } \Omega + 200 \text{ } \Omega} - \frac{400 \text{ } \Omega}{400 \text{ } \Omega + 400 \text{ } \Omega} \right)$$

$$V_M = 2.4 \text{ V}$$

Example 4.7 Design an active low pass filter with gain of 5 and a cutoff frequency $f_c = 2000$ Hz.

Solution

Active filters use op-amps to enhance performance. Figure 4.21 shows an op-amp with a basic R–C low pass filter connected to the non-inverting terminal. The gain is defined by Eq. (4.14)

$$V_{out} = \left(1 + \frac{R_f}{R_g}\right)V_{in} \Rightarrow gain = 5 = 1 + \frac{R_f}{R_g}$$

Therefore,

$$\frac{R_f}{R_g} = 4$$

Selecting $R_f = 8$ kΩ, R_g will take the value 2 kΩ. The filter resistor R can also be chosen as $R = 2$ kΩ. The cutoff frequency Eq. (4.23) can be used to determine the filter capacitor C,

$$f_c = \frac{1}{2\pi RC} \Rightarrow C = \frac{1}{2\pi Rf_c}$$

$$C = \frac{1}{2\pi(2000\Omega)(2000H_z)} = 0.04 \ \mu F$$

Example 4.8 Design a signal conditioning circuit using op-amps to convert the output from a sensor [− 10 to + 10 V] output range from a sensor into a [0 to + 5 V] which can be accepted by a data acquisition system (DAQ).

Solution

Assume a linear relationship between the inputs and outputs of the signal conditioning (Fig. 4.31).

$$V_{out} = AV_{in} + b$$
$$0 = A(-10 \text{ V}) + b$$

Fig. 4.31 Signal flow diagram for Example 4.8

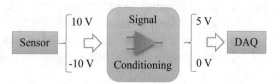

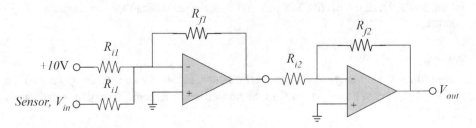

Fig. 4.32 Suggested circuit for Example 4.8

$$5 = A(+10\,\text{V}) + b$$

$$\text{Solving,} \quad \Rightarrow A = 0.25 \text{ and } b = 2.5 \text{ V}$$

The signal conditioning circuit relation is.
$V_{out} = 2.5V_{in} + 2.5$, or $V_{out} = 0.25(V_{in} + 10 \text{ V})$.

This can be achieved by a summing op-amp with two inputs. One input is fixed at a constant value of 10 V and the other input accepts the signal output which can take any value between $-$ 10 and $+10$ V. An inverting op-amp with gain ($K = -1$) is used at the output of the summer to ensure positive output since the summer has a negative gain. The circuit of the signal conditioning circuit is shown in Fig. 4.32. A voltage follower could be placed between the sensor and Ri1 to if impedance matching is not satisfied between the sensor and the input the summer op-amp.

Example values for the resistors in the circuit to satisfy the signal conditioning requirement are.

$R_{i1} = 200 \ \Omega$, $R_{f1} = 50 \ \Omega$.
$R_{i2} = R_{f2} = 50 \ \Omega$.

4.10   Problems

4.1. Calculate and plot the output V_{out} of a derivative op-amp (shown in Fig. 4.14) for the following input signal (Fig. 4.33). Use $R_f = 150$ kΩ, and $C = 12$ μF.
4.2. Design an inverting Schmitt trigger in which and input signal V_{in} needs to decline below 1.5 V for V_{out} to switch to $+ 5$ V. and V_{in} needs to rise above 3.5 V for V_{out} to switch to 0 V. (Hint, you may need to make assumptions on the magnitude of at least one resistor for this Schmitt trigger circuit).

Fig. 4.33 Input signal for Problem 4.1

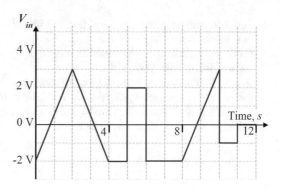

4.3. A Wheatstone bridge is powered by a 12 V DC battery. It is designed with the following four resistors.

$R_1 = 300\ \Omega,\ R_2 = 200\ \Omega,\ R_3 = 200\ \Omega,$ and $R_4 = 300\ \Omega.$

Determine if the bridge is balanced and calculate the power dissipation in each resistor.

4.4. A transducer full scale output ranges from 0 to 10 V is to be rescale to the range 0 to 5 V so it can be accepted as input to a data acquisition system. Design a signal conditioning circuit to achieve this using op-amps without changing the polarity of the sensor signal.

4.5. The resistors in a Wheatstone bridge (Fig. 4.5) are given as

$R_1 = 120\ \Omega,\ R_2 = 200\ \Omega,\ R_3 = 120\ \Omega,$ and $R_4 = 205\ \Omega.$

If the bridge is supplied with $V_s = 24.0$ V, determine the voltage offset V_M.

4.6. Design a high impedance amplifier with gain of 34. Use a non-inverting op-amp and specify all resistors.

4.7. Current is preferred to transmit data instead of voltage to avoid errors associated with loading effects. A voltage-to-current converter (or transmitter) is depicted in Fig. 4.17. Use this as the basis to design a transmitter for a sensor signal output (0–2 V) to become (0–20 mA).

4.8. Design a high input impedance integrator op-amp circuit to generate a linear ramp voltage rising at a rate of 5 V per millisecond.

References

1. R. Prasad, Analog and Digital Electronic Circuits Fundamentals, Analysis, and Applications, Springer Nature, Switzerland AG, 2021. https://doi.org/10.1007/978-3-030-65129-9
2. Basis, Peter. Introduction to electronics: a basic approach. Pearson, Boston, 2014

3. James M. Fiore, Operational Amplifiers & Linear Integrated Circuits: Theory and Application, 3rd ed., dissidents, 2018.
4. Faulkenberry, Luces M., An introduction to operational amplifiers, with linear IC applications. 2nd ed., Wiley, New York, 1982.
5. John V. Wait, Lawrence P. Huelsman, and Granino A. Korn, Introduction to operational amplifier theory and applications. McGraw-Hill, New York, 1975.
6. Chen, Wai-Kai (Editor), The circuits and filters handbook: Fundamentals of circuits and filters. 3rd ed., CRC Press, Boca Raton, 2009.

Semiconductors and Logic Circuits

5

5.1 Introduction

This chapter is intended to build a basic foundation that focuses on the functional mode for circuits commonly used in semiconductors and logic devices. Semiconductors are solid materials that can behave as either conductors or insulators under specific electrical, thermal, or light operating conditions. Examples of semiconductor materials used in the fabrication of semiconductor devices are silicon (Si) and germanium (Ge). These materials are not very useful in their pure state. Useful electronic properties are produced by a controlled process called 'doping.' In this process a small number of foreign atoms, such as boron (b) and phosphorus (P), are added to the pure crystalline structure. For example, when doping silicon with boron atoms (3 valence electrons), one of the four silicon valence electrons from a silicon atom will not be in a bond with a corresponding boron atom creating a positive charge or 'hole'. This positive charge can be filled by an electron from nearby silicon atom leaving behind another hole. This conduction activity of the electrons and holes can be sustained within the crystal structure under the influence of an externally imposed electric voltage or current. The material in this case is referred to as a ***p-type semiconductor*** (Fig. 5.1a). On the other hand, when doping silicon (4 valence electrons) with phosphorous (5 valence electrons), four covalent bonds are formed between a phosphorous atom and adjacent silicon atoms leaving a fifth electron free to migrate through the crystal lattice under an externally applied electric field. Because of these unbonded 'free' electrons, this type of material is called ***n-type semiconductor*** (Fig. 5.1b).

Two or more layers of these n-type and p-type semiconductor materials can be joined in different configurations under controlled manufacturing conditions to form junctions of semiconductor devices that have significant importance. In the following we will discuss some of these semiconductor devices. For more on the topics covered in this chapter, see [1–4].

© The Author(s), under exclusive license to Springer Nature Switzerland AG 2022
I. Abu-Mahfouz, *Instrumentation: Theory and Practice, Part 1*, Synthesis Lectures on Mechanical Engineering, https://doi.org/10.1007/978-3-031-15246-7_5

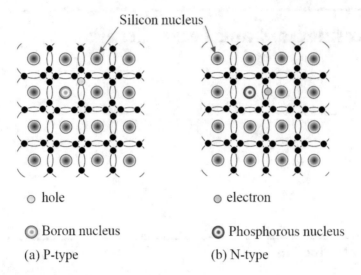

Fig. 5.1 Semiconductor silicon produced by doping with impurity elements with 3 (boron) and 5 (phosphorous) valence electrons

5.2   The Diode

A diode is a simple two-layers (p–n junction) semiconductor device. As shown in Fig. 5.2a, it is characterized by allowing current to flow only in one direction (forward biased). The current flow increases rapidly once the V_{diode} exceeds the forward bias threshold V_f (0.6–0.7 V for silicon diodes). Diodes are not to be used in the reverse bias region since exceeding the reverse-breakdown voltage V_r is destructive to the diode. A Zenner diode is a special type of diodes that allows conduction in the reverse direction without self-destruction.

Other types of diodes include the light emitting diode (LED) and the photo-diode. LED's emit light when they are switched into conduction mode. They are similar to the silicon diodes but have higher forward voltage V_f that varies with the color of the emitted light. Photo-diodes operate opposite to LED's in that when they sense light they conduct and allow current to flow in proportion to the intensity of the received light.

5.3   Bipolar Junction Transistors

The bipolar junction transistor (BJT) is a solid-state device that has three terminals: the base, emitter, and collector. It is usually connected in series with the load. There are two basic types of BJT transistors: NPN and PNP, shown in Fig. 5.3. This transistor can be considered as a current amplifier with the forward gain β defined as;

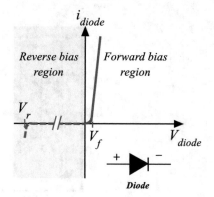

(a) Voltage-Current characteristic for a diode.

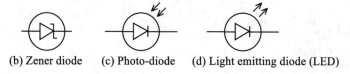

(b) Zener diode   (c) Photo-diode   (d) Light emitting diode (LED)

Fig. 5.2 **a** Characteristic response curve of diode. **b–d** different types of diodes

Fig. 5.3 NPN and PNP transistors (BJT)

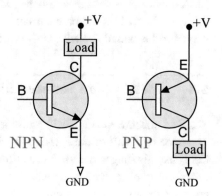

$$\beta = \frac{I_C}{I_B} \tag{5.1}$$

The current equation for the BJT is

$$I_E = I_C + I_B \tag{5.2}$$

with $I_B \ll I_C$, it can be stated that $I_E \approx I_C$

where

$\beta =$ BJT current gain

Fig. 5.4 BJT characteristic
curve

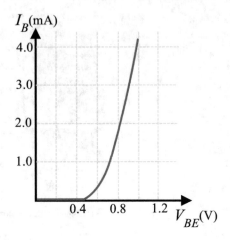

I_C = Collector current
I_B = Base current
I_E = Emitter current

The collector-emitter current I_C is controlled (amplified) by the small current at the base
I_B. A forward-bias voltage across the base-emitter junction, with a minimum magnitude of
0.6 V, is required for the base current to start flowing in a silicon BJT. The characteristic
curve for this biased behavior is shown in Fig. 5.4.

5.4   Metal-Oxide Semiconductor Field Effect Transistors

The field effect transistor (FET) is also a three-terminal semiconductor amplifying device.
The three terminals of the FET are the drain (D), the source (S), and the gate (G). FETs
has some advantages over BJTs, and therefor are more commonly used in power applications. These advantages include high input impedance, high switching speeds, high
current capacity, and less temperature sensitivity. There are several types of field effect
transistors and only the metal-oxide semiconductor FET (MOSFET) is introduced here.
The gate is capacitively coupled to the other terminals leading to a high input impedance
of this transistor. The MOSFET operates in the enhancement mode in which the gate voltage is always positive. Figure 5.5 depicts the MOSFET symbol and its switching mode
characteristic curve.

For all types of FETs the input control signal is the gate-source voltage V_{GS} and the
output is the drain-source current I_{DS}. The gain of a FET (gm) is determined by Eq. (5.3)
and is called the transconductance.

$$gm = \frac{\Delta I_{DS}}{\Delta V_{GS}} \; S(siemens) \, or, \; \text{Amps/Volt} \tag{5.3}$$

Fig. 5.5 The metal oxide semiconductor field effect transistor

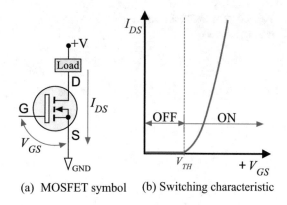

(a) MOSFET symbol   (b) Switching characteristic

where,

gm = FET gain (transcondcutance)

I_{DS} = current from drain to source

V_{GS} = Gate–source input control voltage

For a gate voltage $V_{GS} = 0$ and up to just below the threshold voltage V_{TH}, no load current flows through the MOSFET transistors ($I_{DS} = 0$) and the MOSFET is "OFF" or is in a nonconduction mode. At $V_{GS} = V_{TH}$ the MOSFET starts to conduct and turns "ON". As V_{GS} is increased above V_{TH} the MOSFET resistance decreases, and the drain current I_{DS} increases as presented by the switching characteristic curve in Fig. 5.5b. The gate is electrically isolated from the drain-source channel which gives the enhancement MOSFET its high input impedance characteristic.

Figure 5.6 shows a set of output characteristics curves for N-enhancement mode MOSFET where the drain current I_{DS} is plotted against the drain-source voltage V_{DS} for increasing values of V_{GS}. The ohmic or active region is defined for a small levels of V_{DS} where for a given $V_{GS} > V_{TH}$, the I_{DS} increases with increasing V_{DS}. For sufficiently high values of V_{DS} the I_{DS} levels off and while increasing with V_{GS} it is Independent of V_{DS}. There is an upper limit value of V_{GS} at which the MOSFET current I_{DS} reaches its maximum magnitude I_{DSS} (saturation).

Integrated Circuits (IC) manufacturing is mostly based on two types of technologies: the Transistor–Transistor Logic (TTL) and the Complementary Metal Oxide Semiconductor (CMOS). TTL IC circuits are made up of multiple bipolar junction transistors (BJTs). Examples of TTL ICs include logic gates. On the other hand, CMOS ICs are mostly based on both p-type and n-type MOSFETs to perform digital logic circuits, microprocessors, signal converters, sensors, memory chips. In general, TTL chips consume more power when compared to the power consumed by equivalent CMOS chips. The clock rate is one of the major factors for power consumption. Higher clock values will result in higher power consumption. However, CMOS chips are very sensitive to electrostatic discharge and electromagnetic disruptions, therefore they are more delicate to handle. A

Fig. 5.6 The MOSFET output
characteristics

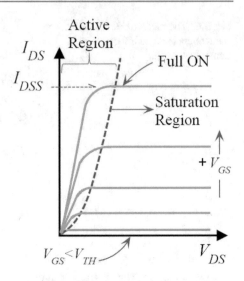

very minute amount of static electricity could cause damage to the CMOS chips. The
fan-out number which indicates the number of standard loads that could be connected to
the output of the gate under the normal operation, is much higher for the CMOS than
it is for the TTL. Furthermore, CMOS circuits are faster in signal transmission and are
smaller in size when compared to TTL circuits for equivalent application.

5.5   Digital Logic

The study of Digital logic is fundamental to the understanding of the operating principles
of ICs, microprocessors, and devices used in instrumentation, sensors, and data acquisition
systems. Some of these chips are made of large networks of logic gates to provide for
a particular function. Logic gates are the basic building blocks of digital electronics and
microprocessors. Logic gates are made of transistors (such as BJTs and MOSFETs) that
process data in discrete form. Digital circuits are less susceptible to noise and are more
reliable for information processing than analog circuits. Digital logic circuits are generally
classified into two groups: combinational logic circuits and sequential logic circuits.

5.5.1   Combinational Logic Circuits

The output of combinational logic at any instant in time is determined by the current state
of its inputs and is independent of the previous (historical) states of its inputs or outputs.
A schematic of two circuits is shown in Fig. 5.7.

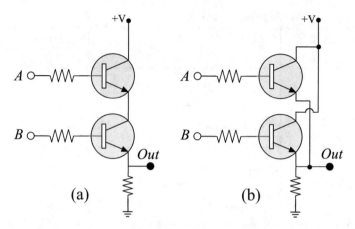

Fig. 5.7 Two logic gate circuits using BJT transistors for **a** AND gate, and **b** OR gate

A sample of basic logic combinational functions, their symbols, and truth table are listed in Table 5.1. The truth table is simply a listing of the gate output for all possible combinations of the inputs. A bar over any variable indicates the inverse or 'NOT' operation for that variable.

5.5.1.1 The Digital Multiplexer (MUX)

Usually, within a DAQ unit, a multiplexer works as a data selector. It scans several input channels and select one channel at a time and forward its value to the ADC through its single common output. The multiplex receives the timing and address of the selected channel from the microprocessor over the select lines. Other types of multiplexers can switch inputs channels to multiple outputs but are not pursued in this discussion. A digital multiplexer consists of a circuit of high speed combinational logic gates. Figure 5.8 shows a schematic diagram for a two-input channel selector circuit, which selects one of two input channels to be connected to the output. In this example, a high (1) signal on the select control line will transmit the input at channel 1 to the output, while a low (0) on the selector line will transmit the input at channel 0 to the output. The schematic of a four-input channels multiplexer is shown in Fig. 5.9. In this case it is clear that two select controls are needed. In general, a multiplexer with 2^n input channels require n select controls.

5.5.2 Sequential Logic Circuits

Sequential logic circuits make use of the past states of their inputs (memory) in addition to the current input states due to a feedback loop from the output to the input. These circuits are also called bistable or two-state devices because their output can take one of

Table 5.1 Basic logic combinational functions

Gate name function	Symbol	Truth table		
		A	B	C
AND $C = A \cdot B$		0	0	0
		1	0	0
		0	1	0
		1	1	1
NAND $C = \overline{A \cdot B}$		0	0	1
		1	0	1
		0	1	1
		1	1	0
OR $C = A + B$		0	0	0
		1	0	1
		0	1	1
		1	1	1
NOR $C = \overline{A + B}$		0	0	1
		1	0	0
		0	1	0
		1	1	0
Buffer $C = A$		–	–	–
		0	–	0
		1	–	1
		–	–	–
NOT (Inverter) $C = \overline{A}$		–	–	–
		0	–	1
		1	–	0
		–	–	–

Fig. 5.8 Two-input channel selector (MUX)

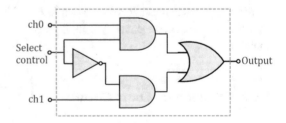

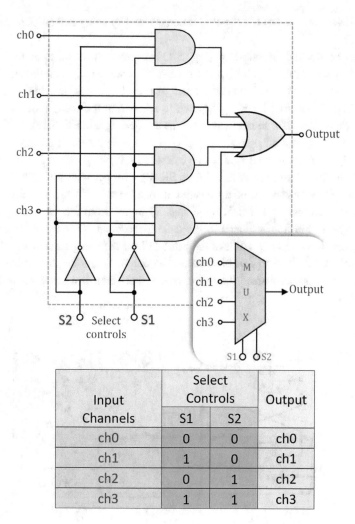

Fig. 5.9 Four-input channel selector (MUX)

Input Channels	Select Controls		Output
	S1	S2	
ch0	0	0	ch0
ch1	1	0	ch1
ch2	0	1	ch2
ch3	1	1	ch3

two logic states, either 1 or 0. The output is kept 'latched' at its current state until the device receives a trigger signal or clock pulse which changes the output state.

5.5.2.1 Flip-Flop Circuits

A simple logic device that demonstrates the sequential dependence of events is the SR (Set-Reset) flip-flop circuit. This device has two inputs and latches the output Q at one of the binary states until the next change event of its inputs. The $\overline{Q}$ output is the complement of the Q output. The device will cycle through the feedback loop until it reaches a steady state condition. One type of SR flip-flop is constructed by using two NOR gates as shown in Fig. 5.10.

A gated or clocked SR flip-flop (Fig. 5.11) responds to changes at its inputs only when the 'Enable' or clock (CLK) input is high '1'. Otherwise, when the enable input is low '0', the outputs Q and $\overline{Q}$ remain latched at their last states. In addition to the latch type, flip-flop circuits can be edge-triggered to respond at the rising (positive) or falling (negative) transitions of a square-wave such as a clock signal (Fig. 5.12).

The JK flip-flop is similar to the SR (J = S and K = R) but it allows the condition when both J and K are at level 1, in which case the JK flip-flop output toggles (Fig. 5.13).

The D flip-flop is equivalent to a JK or SR flip-flop but with a single 'Data' or 'D' line that is split with the use of an inverter in one branch as shown in Fig. 5.14. This modification prevents the J and K (or S and R) inputs from having the same logic at one time. The D flip-flop responds to the data at the input 'D' only at clock transition (edge-trigger). The output Q will take the value of D at the rising edge of the clock pulse. When the clock line is at low or high states, the output Q and its complement $\overline{Q}$ do not change their values, but instead they store 'remember' their last states regardless of the logic level at D.

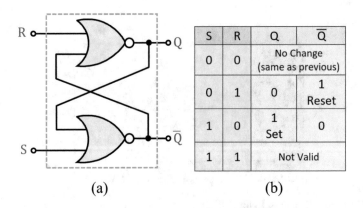

S	R	Q	$\overline{Q}$
0	0	No Change (same as previous)	
0	1	0	1 Reset
1	0	1 Set	0
1	1	Not Valid	

(a)   (b)

Fig. 5.10 SR flip-flop; **a** equivalent NOR circuit, and **b** truth table

Fig. 5.11 Latch SR flip-flop

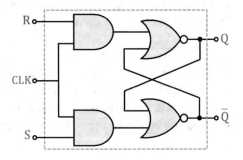

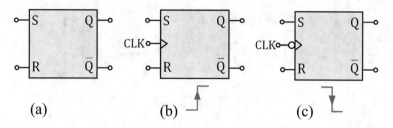

Fig. 5.12 SR flip-flops; **a** transparent, **b** rising edge-triggered, and **c** falling edge-triggered

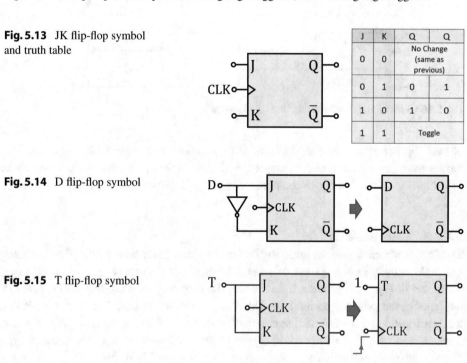

Fig. 5.13 JK flip-flop symbol and truth table

J	K	Q	Q
0	0	No Change (same as previous)	
0	1	0	1
1	0	1	0
1	1	Toggle	

Fig. 5.14 D flip-flop symbol

Fig. 5.15 T flip-flop symbol

D flip-flops are commonly used as data registers and as output latches in digital-to-analog DAC circuits. A Toggle 'T' flip-flop is a single bit bistable logic circuit that is created by connecting and setting both the J and K data inputs to level high or '1' in a basic JK flip-flop as shown in Fig. 5.15. The outputs Q and invert their previous states at each rising edge of the clock pulse. T flip-flops are used in binary counters, frequency dividers, and in memory storage devices.

5.5.2.2 Digital Counter

Figure 5.16 shows the circuit and signals for a 3-bit binary counter using three T flip-flops. The clock signal is applied to the input of the first flip-flop on the left. The output

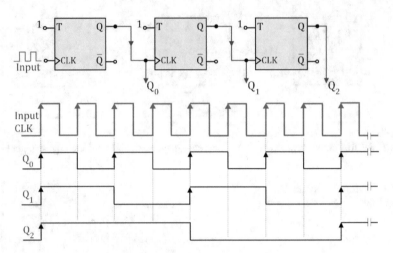

Fig. 5.16 3-bit digital counter using three T flip-flops

of each flip-flop is also used as the input of the next one to its right in the sequence. The arrows show the sequence of the rising edge-triggers as they propagate through this chain of T flip-flops. This is a ripple counter in which the Q outputs do not toggle at the same time.

5.5.2.3 Shift Register

The Shift Registers are devices that are used to store or transfer binary data. These sequential logic circuits are composed of several D flip-flops connected in a serial chain as illustrated in Fig. 5.17. Starting from the left, the output of each flip-flop is connected to the input of the next one on its right. The data present on the Data line is first loaded to the first flip-flop on the left, and then is moved or 'shifted' to the right one flip-flop at a time, at each rising (or falling edge) trigger of the clock input signal. All cascaded flip-flops in one device share a single clock signal. This means that all flip-flops in the device will shift their stored data simultaneously. A single shift register device is constructed from a number of flip-flops equal to the number of bits to be stored. Data can be shifted in or out in a serial (left or right) direction or in a parallel (up or down) configuration.

Mixed data shift modes of serial (in or out) with parallel (in or out) are also possible. Four types of 4-bit shift registers are schematically presented in Fig. 5.17 for comparison. With the exception of the PIPO shift register, which directly presents the 4 input data bits at the 4 Q outputs at a single clock trigger, the other three registers require 4 clock pulses to complete the shifting of the 4 input data bits through the 4 flip-flops. The PISO (Fig. 5.17d) the selector line (shift/load) start by activating the load which, by using a simple combination multiplexer device, allows all 4 input bits (D0, D1, D2 and D3) to be registered (stored) in their respective flip-flop devices. Then the shift mode of the selector

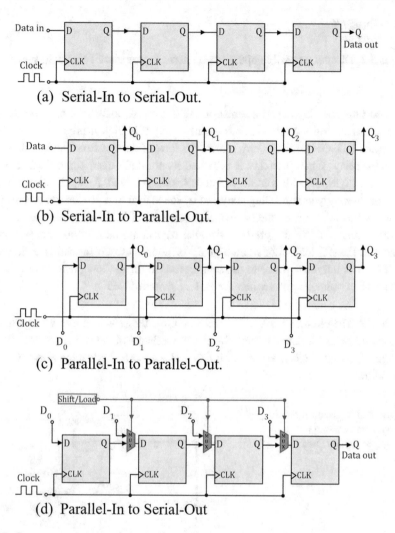

(a) Serial-In to Serial-Out.

(b) Serial-In to Parallel-Out.

(c) Parallel-In to Parallel-Out.

(d) Parallel-In to Serial-Out

Fig. 5.17   Four types of shift registers; **a** SISO, **b** SIPO, **c** PIPO, and **d** PISO

will allow the output of each flip-flop to be shifted (transmitted) to the next flip-flip on the right at the clock trigger pulse (serial fashion). Four clock pulses are required to complete the serial data transmission to the data output bit on the right before another load operation start again. A universal shift register, such as the TTL 74LS194, is a bi-directional multifunctional data register capable of performing different modes of data shifting, storage, delay, and data transfer in serial, parallel or mixed formats. They are also used in calculators and computers to facilitate arithmetic operations.

5.5.3   Examples

Example 5.1 Discuss a possible operating scenario for the circuit shown in Fig. 5.18.

Solution
The circuit illustrates the use of a sensor to control an AC load, such as a motor, through using a transistor and a relay. When the sensor output is high it provides a base current to switch the BJT ON which in turn allows the flow of current from the + Vdc source to energize the relay. When the relay is activated its normally open contact will close. This will close the high-power AC circuit which drives the AC load. An application could be an position or proximity sensor being activated by the arrival of a product on a conveyor belt at a specified location which then activated a robot arm (AC load) to pick up the product or perform some task for the product. Another case could be a sensor that will turn ON the power to the AC load only when a guard or door is closed for safety condition. The diode is used as a flyback diode to provide protection for the low power transistor circuit from current or voltage spikes when the BJT is switched OFF.

Example 5.2 The circuit shown in Fig. 5.19 controls the sensor output voltage level V_{out} using a transistor in the sensor circuit. Specify the resistors of the voltage divider R_1 and R_2 so that the $V_{out} = 0$ V when the sensor is ON and $V_{out} = 5$ V when the sensor is ON. Use Vdc $= 30$ V.

Fig. 5.18 A BJT circuit for AC motor control based on sensor output signal (Example 5.1)

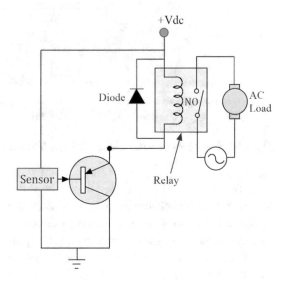

Fig. 5.19 A sensor output V_{out} is controlled by a BJT circuit for Example 5.2

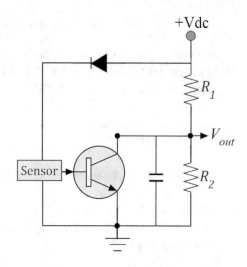

Solution

When the sensor is ON it provides a base current switching the transistor into saturation state (fully ON). At this state maximum current will flow through R_1 and the voltage drop across the BJT will be very small (0.2 V). $V_{out} \approx 0.2$ V or ≈ 0 V. The current flow through R_1 can be determined by

$$I_{CE} = \frac{30 \text{ V} - 0.2 \text{ V}}{1000 \ \Omega} = 0.0298 \text{ A} = 29.8 \text{ mA}$$

When the sensor is OFF the transistor will be open and $I_{CE} = 0$. The current will flow from the source Vdc through the voltage divider circuit.

Selecting $R_1 = 1$ kΩ, R_2 can be determined to give $V_{out} = 5$ V

$$V_{out} = Vdc\left(\frac{R_2}{R_1 + R_2}\right)$$

$$5 \text{ V} = 30 \text{ V}\left(\frac{R_2}{1000 \ \Omega + R_2}\right) \ \Rightarrow \ R_2 = 200 \ \Omega$$

Example 5.3 For the logic MOSFET shown in Fig. 5.20 if the resistance of the MOSFET is 0.1 Ω when it is ON, discuss the operation mode of this transistor.

Solution

A V_{GS} of 4 V is sufficient to turn ON the MOSFET. Calculate the drain current.

$$I_{DS} = \frac{12 \text{ V}}{1000.1 \ \Omega} = 0.012 \text{ A} = 12 \text{ mA}$$

This will give a voltage drop across the MOSFET of magnitude

$$V_{DS} = 0.012 \text{ A} \times 0.1 \ \Omega = 1.2 \text{ mV}$$

This small value of VDS is well in the linear region of MOSFET operation (Fig. 5.6). Reducing the load resistance to 100 Ω and recalculating gives $I_{DS} = 120$ mA and $V_{DS} = 12$ mV which is low and the MOSFET is still in the linear range.

5.6   Problems

5.1. Determine the load current I_C and the voltage drop V_{CE} for the BJT transistor circuit shown in Fig. 5.21.
5.2. For the BJT Circuit Shown in Fig. 5.22. Use the Given Data to Determine the Voltages at VC and VE. Is the Transistor Operating in the Linear Region or in the Saturation Region?

Fig. 5.20 BJT circuit for Example 5.3

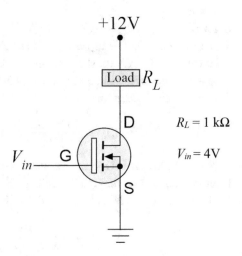

Fig. 5.21 BJT circuit for Problem 5.1

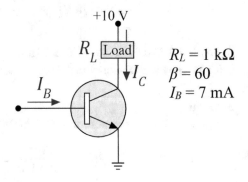

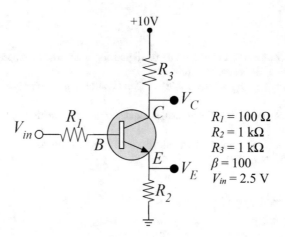

Fig. 5.22 BJT circuit for Problem 5.2

5.3. A BJT transistor with a gain of $\beta = 70$ and operating within its linear region is to operate with a load current I_C of 4 A. Determine the required base current I_B.

5.4. Draw a combinational logic circuit to realize the relation.

$$Q = A \cdot C + \overline{A} \cdot B + A \cdot B \cdot \overline{C}$$

5.5. Draw the Output (Q) Diagram Shown in Fig. 5.23 for a Positive Edge Triggered JK Flip-Flop.

5.6. Establish a Comparison Table and List the Differences Between SR, JK, D, and T Flip-Flop Circuits.

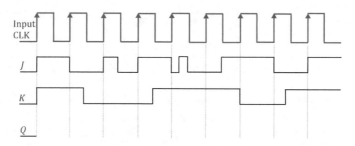

Fig. 5.23 Input signals at CLK, J, and K for Problem 5.5

References

1. Rafiquzzaman, Mohamed, Fundamentals of digital logic and microcomputer design. 5th ed., Wiley-Interscience, Hoboken, N.J. 2005.
2. Tertulien Ndjountche, Digital electronics 1: Combinational logic circuits. Wiley, New Jersey, 2016.
3. Tertulien Ndjountche, Digital electronics 2: sequential and arithmetic logic circuits Wiley, New Jersey, 2016.
4. Coughlin, Robert F., Principles and applications of semiconductors and circuits. Prentice-Hall, Englewood Cliffs, N.J. 1971.

Data Acquisition Systems

6

6.1 Introduction

To measure physical variables like temperature, pressure, force, displacement, light, fluid flow velocity, and others, electrical or electromechanical devices are used to convert these variables into voltage or current signals. These devices are called sensors or transducers. Commonly used sensors and transducers are discussed in Part II of this book [1]. The output signals from these transducers are usually continuous (i.e., analog) in magnitude and time. These measurement signals are used as inputs to the data acquisition system (DAQ or DAS). Data acquisition systems are at the core of all devices that require the input of a measured physical or process variable. The acquired data is usually sent to a microprocessor for processing and analysis. This data could be part of a machine condition monitoring setup or used as feedback information in closed loop automatic control applications. While most sensors and transducers are composed of analog circuits that process analog signals, DAQ systems and microprocessor are predominantly designed and made using digital electronic components. All modern electrical devices are based on digital (binary) data. Therefore, analog signals must be converted to digital signals before they can be used in any modern application. Digital circuits process discrete signals which vary between two levels of voltages. Generally, these two discrete levels are either ON (some positive level of voltage like 5 V) or OFF (0 V). In binary format these are either 1 or 0. Unlike analog signals, digital signals are less susceptible to noise and can be easily stored and processed using modern computer technology.

As discussed in Chap. 5, digital circuits are essentially constructed using simple logic gates, inverters, and flip-flop devices. Digital circuits are composed of fast switching devices like diodes, transistors, multiplexers, combinational logic and sequential logic circuits [2]. The goal of the discussion in this chapter is to discuss the working principles of common circuits used in data acquisition systems.

© The Author(s), under exclusive license to Springer Nature Switzerland AG 2022   115
I. Abu-Mahfouz, *Instrumentation: Theory and Practice, Part 1*, Synthesis Lectures
on Mechanical Engineering, https://doi.org/10.1007/978-3-031-15246-7_6

6.2   Data Acquisition System (DAQ)

Modern data collection is based on the use of microprocessors or microcontroller units (MCU) to manage and execute the many tasks that are required for acquiring measurements [3]. Figure 6.1 shows a schematical presentation of a generalized measurement system. The tasks of taking a measurement may include, sensing the PV, signal conditioning, signal transmission, analog to digital conversion (ADC), storage, processing/analysis, and interpretation. The power of programming DAQ systems provide the user great flexibility and easy access to control the measurement operations such as selecting input channels (sensor), sampling rate, duration of data acquisition, storage, and data presentation. Interface between the DAQ and other instruments is facilitated via serial I/O ports, ethernet, wireless communication protocols. Wide variety of DAQ systems are available to choose from. For example, a Digital storage Oscilloscope (DSO) is a stand-alone device that is capable of acquiring, processing, displaying and storing signals. The DSO can transmit data to other devices such as computers for further processing and analysis.

Data Loggers are examples of stand-alone portable DAQ devices that are used to record data over a designated period of time. These are self-contained data acquisition systems with a built-in processor, embedded software, and with local storage memory.

As stated earlier, analog signals produce by sensor during measurements must be converted to digital numbers using an analog to digital converter (ADC). At the output from a microcontroller, a digital to analog converter (DAC) convert each value of the digital control signal into a corresponding analog output signal that can be utilized by an actuator.

The microcontroller operates in a discrete fashion where at a certain instant in time it is busy manipulating a single sampled representation of each one of the input signals to

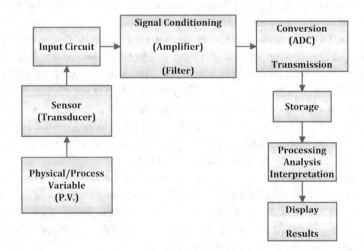

Fig. 6.1 A block diagram of a generalized measurement system

produce a control output signal. Therefore, as the microcontroller is limited by the speed (frequency) of its clock, the time to execute program statement, and the time to convert data at both input (ADC) and out (DAC) stages only discrete (sampled) information from the original continuous analog signal (produced by sensor or input devices) are available for use. All other information is lost and are not available to the microprocessor.

During data acquisition, the tasks of the DAQ are automatically repeated in a cyclic manner to respond to changing states at its input and output connections. During each sweep, the MCU, in the DAQ, reads the states at its inputs, perform programmed mathematical and logic operations (execute program), and updates the outputs. The total sweep time is therefore composed of sampling time, conversion time, and program execution time in addition to other times such as data transfer time. It is therefore important for the experimenter to understand and account for these times to achieve a successful interface between sensors, DAQ systems, controller, and actuators (as, for example, in an industrial automation application).

DAQ systems with variety of configurations and speeds are available. These devices can have a single or multiple channels for analog inputs/outputs and digital inputs/outputs. They are also available with a fixed or modular platform. In the modular type, the measurement system can be customized to meet exact purposes but allow for future change and expansion by choosing from a variety of dedicated input or output modules. The modules can be fitted into a chassis which has the MCU, power supply, and interface ports. In selecting a DAQ system, a good plan regarding the measurement needs is necessary. This plan should consider the types of signals to be measured, acquisition speed (i.e., samples/second), signal generation, acceptable accuracy range, and resolution.

Commonly used DAQ devices are designed to accept signals ranges of ± 5 V or ± 10 V. Most sensors require signal conditioning before their output signals cab be connected to the ADC device. Common signal conditioning tasks include amplification (or attenuation), filtering, and balancing of input circuits (Chap. 4). These can be achieved by using DAQ built-in capabilities or by using external op-amp, filters and other signals conditioning devices.

6.2.1 Resolution and the Quantization Error

Analog signals are continuous, with infinite resolution, in both time and amplitude. This means that, at any instant in time, the analog signal can take any value within its full range. The conversion of a signal from its original analog form into the finals binary code takes several steps. These are sampling, quantization, and encoding. Quantization is the assigning of a discrete value to the sampled analog value out of discrete levels of values defined by the resolution of the ADC which is controlled by the number of bits in the ADC. For an n-bit ADC, there are 2^n discrete states that can be coded into n-bit binary representations. The number of intervals or steps between these states is $2^n - 1$. Hence, it

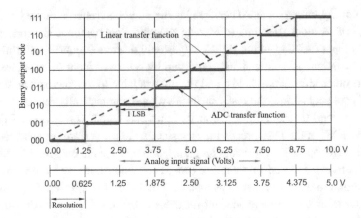

Fig. 6.2 ADC transfer function showing two analog input examples; $V_{FSR} = 5$ V and $V_{FSR} = 10$ V

can be stated that the resolution is the minimum change in the input analog signal that can be detected (converted) by the ADC. The conversion resolution can be calculated using the following relation

$$Resolution = \frac{Voltage\,(Full\,scale\,Range)}{2^n} = \frac{V_{FSR}}{2^n} \tag{6.1}$$

where the full scale range is the maximum voltage range that can be converted by the ADC. It is generally equivalent to the reference voltage V_{ref} of the ADC. For a fixed V_{FSR} or V_{ref}, the resolution can be enhanced by increasing the number of bits for an ADC. Figure 6.2 shows the transfer function for a 3-bit ADC. Input signals to the ADC outside the Full-scale range will cause the ADC to saturate at one of its limits (i.e., 000 below the minimum V_{in} and 111 above the maximum V_{in}). Therefore, the smallest change in the measured quantity that can be recognized by the DAQ microprocessor is equivalent to the value calculated by Eq. (6.1).

During conversion some information in the analog signals will be lost. As can be clearly seen, due to the quantization process an error is introduced in the output of the ADC. This error is numerically equal to one conversation step or equal to the resolution of the ADC and is known as the quantization error. It is also defined in terms of one step or the least significant bit (LSB). As the number of bits used in the converters increases, the resolution, and hence, the quantization error become smaller.

By shifting the axis representing the input signal in Fig. 6.2 to the left by LSB/2 offset, the quantization error is now equal to ± 0.5 LSB (Fig. 6.3) truncated to the lower limit of the step or rounded-off to the upper end of the step.

Quantization and coding take time and the sample value should remain constant until conversion is completed. This is accomplished by the sample and hold circuit.

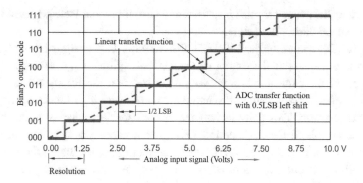

Fig. 6.3 ADC transfer function with ± 0.5 LSB error range for a $V_{FSR} = 10$ V analog input signal

6.2.2 Sample and Hold Circuit

The ADC may produce errors if the input signal varies while a conversion is taking place. Changes of more than half the ADC resolution ($> \pm 1/2$ LSB) at the input voltage of ADC will cause errors in the converted output digital value. As the name indicates, a sample and hold circuit (Fig. 6.4) keeps the input to the ADC at a constant value during the conversion duration until the next sampling cycle.

To understand the working principle of this circuit, Fig. 6.5 illustrates an input and output wave forms for the circuit in Fig. 6.4. The input signal to be sampled (V_{in}) is connected to the drain (D) of a MOSFET that acts as timed ON/OFF switch that charges a capacitor C connected to its source (S). A train of control pulses is applied at the gate (G). A voltage follower connects the potential of the capacitor to the output (V_{out}). Use of other types of op-amps are possible depending on the application and signal conditioning requirements. Another buffer amplifier (not shown) can be connected to buffer the input signal and provide impedance matching. The capacitor C serves as the sampled data

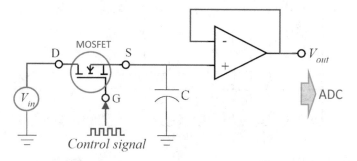

Fig. 6.4 Schematic for a basic sample and hold circuit

storing device. The capacitor must be selected to have minimum leakage possible to allow for a stable signal during hold time.

During the ON trigger of the MOSFET the capacitor C is charged by the input signal V_{in} which also appears at the output of the buffer amplifier V_{out}. When the MOSFET is switched off, the capacitor is not charged further, and ideally should retain the last sampled V_{in} value since it cannot discharge across the very high input impedance of the output buffer op-amp. This constant voltage of the capacitor appears as the V_{out} signal during this hold (MOSFET OFF) period. Other fast triggering devices such as a BJT can also be used in place of the MOSFET. The time duration of switching the transistor ON is very small (in the order of microseconds) and just long enough to charge the capacitor C so that $V_{out} \approx V_{in}$.

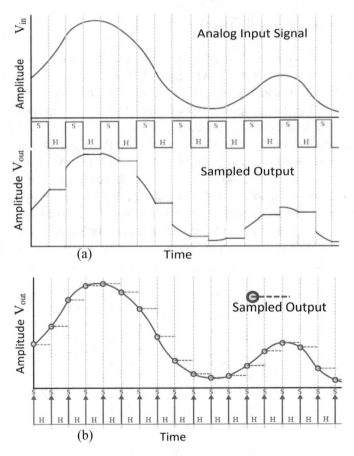

Fig. 6.5 Sample and hold input and output signals, **a** basic concept of S&H circuit signals, **b** ideal form for ADC

6.2.3 Sampling Rate

According to the Nyquist sampling theorem, the minimum sampling rate f_s must be equal to, at least, twice the highest frequency of interest (f_{Nyq}) in the analog input signal

$$\left(f_{Nyq} = \frac{f_s}{2} \right). \tag{6.2}$$

Therefore, f_{Nyq} is the highest frequency that can be realized at a given sampling rate f_s. Dynamic components within an acquired signal having frequencies greater than f_{Nyq} are not accurately represented. For a uniformly sampled signal the time interval between samples is Δt_s and the sampling rate, or sampling frequency, is $f_s = 1/\Delta t_s$ samples per second (s/s) or (Hz). At the ith sampling instant t_i

$$t_i = i \Delta t_s, \text{ and } x(t_i) = x(i \Delta t_s)$$

where,

x_i is the ith sample value of $x(t)$ at discrete-time instant t_i.

Figure 6.6 illustrates the effect of varying the sampling rate f_s on the output signal obtained by discrete sampling. The more samples that can be acquired within a certain period the more accurately the original signal can be reconstructed. Aliasing effects occurs in the constructed (acquired) signal if the sampling rate is less than twice the maximum frequency of the original signal. However, in practice, the sampling rate should be taken as at least 5 times or 10 times the maximum frequency of interest to ensure a more accurate representation of the shape of the original sensor output signal.

6.3 Digital-To-Analog Converters

A digital to analog converter (DAC or D/A) is an electronic circuit that converts digital data which exist in binary words or bytes (0's & 1's) format into an analog signal (voltage or current). These devices are usually used at the output of a microcontroller to interface the controller with an analog output device (output). They ae also used in the design of some analog to digital (ADC or A/D) circuits [4, 5].

6.3.1 Summing Amplifier DAC

Figure 6.7 demonstrates a simplified version of a DAC using a summing op-amp with resistor bank of a 4-bit configuration. For a n-bit DAC converter there are ($2^n - 1$) possible levels (steps) of output voltage starting at 0 V. The resolution of the DAC is equivalent to the smallest output voltage step which is also equivalent to the weight of an LSB.

Fig. 6.6 A 5 Hz sine wave (blue color) sampled at different rates f_s with the discrete sampled output shown by connected the data using straight lines (red color)

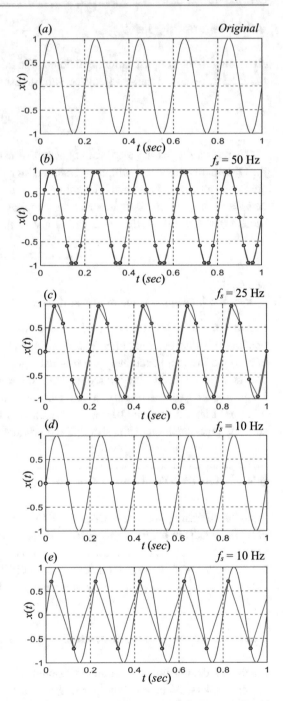

Fig. 6.6 (continued)

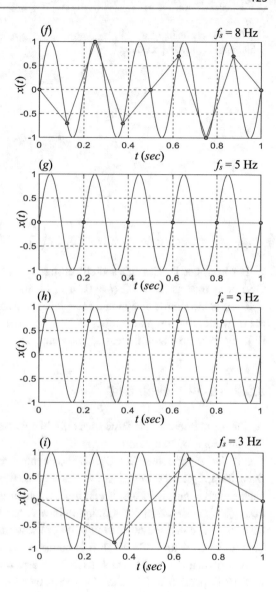

Therefore, higher output resolution requires the use of larger number of bits. The input resistors must be arranged with progressively increasing values by a factor of 2. The input to the op-amp is a 4-bit digital signal that controls the 4 input switches. The switches are BJT or FET type transistor ICs. The desired output voltage V_{out} depends on the selection of the resistor R_f and the reference voltage V_{ref}. Using the gain equation for a summing op-amp, the analog V_{out} is given by

Fig. 6.7 A schematic of a
4-bit summing op-amp DAC

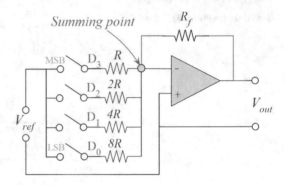

$$V_{out} = -V_{ref}\left(\frac{R_f}{R}\right)\left(\frac{D_3}{1} + \frac{D_2}{2} + \frac{D_1}{4} + \frac{D_0}{8}\right) \tag{6.3}$$

For an n-bot DAC, each switch setting is controlled by the value (0 = OFF or 1 = ON) of its associated D_i bit ($i = 0, 1, 2, \ldots, n$).

Despite the simple design and acceptable conversion speed, this simple weighted resistor network relies on maintaining precise ratio between the resistors. This becomes more difficult to control with increasing bit count.

6.3.2 R-2R Ladder DAC

As the name implies, this DAC uses only two precision resistors R and $2R$ which resolves the challenge of maintaining high accuracy compared to the previous summing DAC. The digital input codes ($D_0, D_1, \ldots, D_n$) control the switches so that only a 1 or 0 state from each digit contributes to the summed analog output V_{out}. As shown in Fig. 6.8, each switch connects its associated input to either ground or to the op-amp virtual ground point (0 V) at the inverting terminal. Node A has two $2R$ resistors connected in parallel (one to ground and the other to virtual 0 V) resulting in an effective resistance of $R_A = R$ between node A and 0 V. Node B is connected to a $2R$ resistor which is in parallel with an R (between B and A) which itself is in series with the effective R at A. This results in a $2R$ in parallel to $2R$ with effective resistance at B of $R_B = R$. This pattern repeats sequentially leading to $R_C = R$ and $R_D = R$. As a result, the current flow in all branches will be as presented in the table depicted within Fig. 6.8 starting with $i_D = V_{ref}/R_D = V_{ref}/R$.

$$V_{out} = V_{ref}\left(\frac{R_f}{R}\right)\left(\frac{D_0}{16} + \frac{D_1}{8} + \frac{D_2}{4} + \frac{D_3}{2}\right)$$

$$V_{out} = V_{ref}\left(\frac{R_f}{R}\right)\left(\frac{D_0 + 2D_1 + 4D_2 + 8D_3}{(2^4 = 16)}\right) \tag{6.4}$$

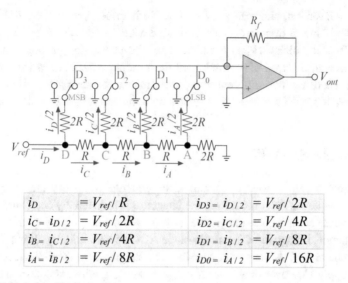

i_D		$= V_{ref}/R$	$i_{D3} = i_{D/2}$	$= V_{ref}/2R$
$i_C = i_{D/2}$		$= V_{ref}/2R$	$i_{D2} = i_{C/2}$	$= V_{ref}/4R$
$i_B = i_{C/2}$		$= V_{ref}/4R$	$i_{D1} = i_{B/2}$	$= V_{ref}/8R$
$i_A = i_{B/2}$		$= V_{ref}/8R$	$i_{D0} = i_{A/2}$	$= V_{ref}/16R$

Fig. 6.8 A 4-bit ladder DAC circuit

The resolution of the 4-bit R-2R converter is equal to the least significant bit given by $V_{LSB} = V_{ref}/2^4 = V_{ref}/16$.

6.4 Analog-To-Digital Converters

An Analog-to-digital converter is a core component in the data acquisition system (DAS). This IC receives signals, at controlled instances of time, from measuring devices such as temperature, displacement, pressure, and sound transducers. The ADC coverts each one of these signals to a discrete level proportional to its input analog value. These discrete quantities are in the form of binary code (digital word) that can be interpreted by the DAQ microprocessor unit. This is in reverse to the operation of the DAC IC discussed in the previous section. The converted discrete data represent the smallest changes in the measured signal that can be recorded by the DAQ system. The higher the number of bits of an ADC the greater the number of discrete levels that can represent a given range of input analog signal and hence the greater is the resolution. For example, an 8-bit ADC can produce a maximum of $2^8 = 256$ discrete levels while a 16-bit ADC produces a maximum of $2^{16} = 65,536$ levels.

The input signal V_{in} must be held at constant value during the conversion process. As discussed in Sect. 6.2.2, this is usually accomplished by a sample-and-hold circuit that controls availability of the analog input signal and its readiness for conversion. A bipolar power supply Vs is required to run the ADC's internal components such as op-amps and digital logic circuits [4, 5]. Single bit control connections, such as start conversion and end of conversion, and read pin facilitate the interface between the ADC and other peripherals within the microprocessor or DAQ device. In addition to the input pin (V_{in}) a stable reference voltage (V_{ref}) is required for the conversion process. The results of conversion for an n-bit ADC are presented on n output parallel lines for connection to a data bus. There are two different approaches in use for analog to digital conversion; integrating ADCs, which utilize an integrator op-amp, and non-integrating ADCs.

6.4.1 Single-Slope ADC

An integrator op-amp is used to incrementally generate a linearly increasing (ramp or slope) output voltage. This output from the integrator is sequentially compared to the input signal V_{in} while counting the integration time (steps) using a binary counter. Figure 6.9 shows a functional diagram of the single-slope ADC device. Initially all input signals are assumed to be at 0 V. At start of conversion, the integrator op-amp integrates a constant reference supply V_{ref} over time, and its output is fed to the inverting terminal of a comparator. The non-inverting terminal of the comparator receives the analog input signal V_{in} to be converted into a digital form. The comparator output will stay high (logical 1) and the counter continue to be incremented by clock pulses until the output from the integrator op-amp reaches a level equal to V_{in} as illustrated in Fig. 6.10. At this point the comparator switches its output to logical 0 leading the 'AND' gate to stop incrementing the counter. The digital output word represents the time of integration which is proportional to the analog input V_{in}.

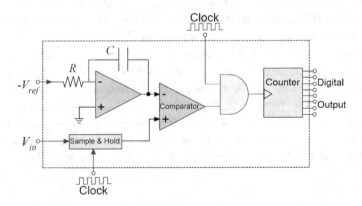

Fig. 6.9 Single-slope ADC

Fig. 6.10 Conversion time for a single-slope ADC. Integration time is proportional to V_{in}, The slope of the ramp is constant with magnitude (V_{ref}/RC)

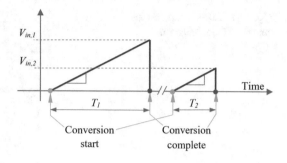

The following relations applies for the integrator op-amp used in this ADC.

$$V_{out}(t) = -\frac{1}{RC}\int_0^t -V_{ref}.dt + \cancel{V_{out}(0)}^0 = \frac{T}{RC}V_{ref}$$

$$V_{out}(t) = V_{in} = \frac{T}{RC}V_{ref} = \left(\frac{V_{ref}}{RC}\right)T \tag{6.5}$$

The single-slope ramp ADC is dependent on the integrator's characteristics defined by R and C which may lead to reduced accuracy and reliability. A more stable ramp ADC is the dual-Slope ADC.

6.4.2 Dual-Slope ADC

In the dual-slope (Fig. 6.11) ADC the input analog voltage V_{in} is allowed to charge the capacitor for a fixed time period T_o. The output of the integrator at this stage is given by

$$V_{out}(T_o) = -\frac{1}{RC}\int_0^{T_o} V_{in}.dt + \cancel{V_{out}(0)}^0 = -\frac{T_o}{RC}V_{in}$$

Next, after T_o has elapsed, the counter is activated and the input to the integrator is switched to accept a reference supply voltage $-V_{ref}$. Integration proceeds until the output of the integrator becomes 0 (Fig. 6.12) and the time interval T_i to achieve this (for each ith analog input $V_{in,i}$) is determined by

Fig. 6.11 A dual-slope digital
converter block diagram

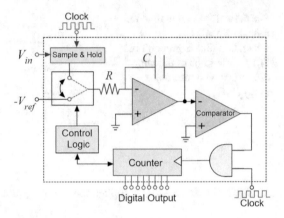

$$V_{out}(T_i) = -\frac{1}{RC} \int_0^{T_i} -V_{ref} \cdot dt + V_{out}(T_o) = 0$$

$$\Rightarrow \quad \frac{T_i}{RC} V_{ref} - \frac{T_o}{RC} V_{in,i} = 0 \tag{6.6}$$

$$\Rightarrow \quad T_i = \left(\frac{T_o}{V_{ref}}\right) V_{in,i}$$

The counter output is a digital word which represents T_i which in turn is directly proportional to the ith analog input $V_{in,i}$ ($T_i \propto V_{in,i}$) and is independent of the integrator's R and C. The total conversion time for $V_{in,i}$ will then be

$$T_{Ci} = T_o + T_i \tag{6.7}$$

Due to their slower speed, ramp ADC's are not commonly used for data acquisition but are used in other measuring devices such as digital multimeters.

Fig. 6.12 Conversion times
and performance
characteristics for a dual-slope
ADC

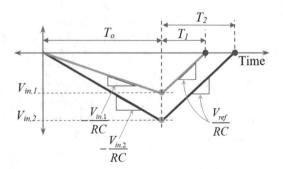

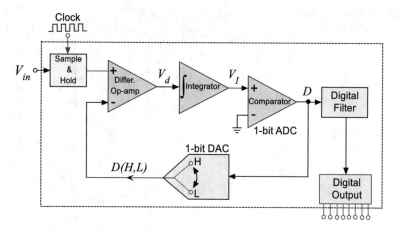

Fig. 6.13 Sigma-delta ADC block diagram

6.4.3 Sigma-Delta ADC

Or Delta-sigma ADC consists of a differential op-amp 'delta', an integrator op-amp 'sigma', and a comparator '1-bit ADC' all connected in series as shown in Fig. 6.13. The output of the comparator is a single bit binary logic that limits the output voltage of a 1 bit-DAC switch to either a high 'H' (if $D = 1$) or a low 'L' (if $D = 0$). This converter accepts high sampling rates and averages the output of the comparator over several clock cycles. The moving average process is performed by the digital filter. The more cycles used in the averaging the more accurate the digital output in representing the input V_{in}. It can reach high resolution and accuracy at reasonable speed. However, the higher the resolution the more cycles required in the moving averaging process and the slower the conversion rate.

6.4.4 Flash ADC

This is a parallel (non-integrating) ADC converter which uses a set of comparator op-amps to compare the input voltage (V_{in}) to a set of reference voltages. This results in a set of digital outputs (0 or 1) that are processed by a priority encoder. Figure 6.14 illustrates the working concept of a 3-bit flash ADC with $2^3 - 1 = 7$ comparators.

Each comparator has its own threshold voltage connected to its inverting terminal. These seven threshold voltages are determined by a string of resistors that form a multiple voltage divider between ground and V_{ref}. As an example, threshold values are presented for $V_{ref} = 8$ V in Fig. 6.14. For a given V_{in}, all comparators with their threshold voltages less than V_{in} will saturate high (output $= 1$) and those with their threshold voltages greater than V_{in} will saturate low (output $= 0$). The logical outputs from all comparators

Fig. 6.14 A 3-bit flash ADC
with 7 comparators

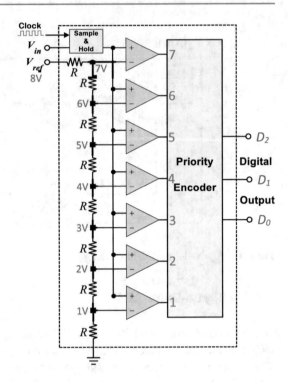

are processed by a priority encoder which interprets the results into the 3-bit digital output
for the converter example shown in Fig. 6.14.

At each clock pulse, all comparators respond to the sampled data V_{in} simultaneously.
Therefore, conversions performed by the flash ADC are very rapid and can reach rates
at the level of giga samples per second. Flash ADCs are also used in large bandwidth
applications such as in radar and satellite communication. However, a major disadvantage
of the flash ADC is that its structure becomes more complex (too many comparators) and
expensive as the number of bits increases for high resolution. These converters require
$2^n - 1$ comparators for an n-bit resolution. In addition, achieving accurate threshold or
voltage references becomes more costly with increasing bit count of this ADC.

Figure 6.15 shows an 8-bit half-flash ADC which consists of two 4-bit flash ADC each
operating on half of the final 8-bit resolution. The result of the first 4-bit flash ADC forms
the 4 MSB of the results and is also converted on its analog equivalence using a 4-bit
DAC. Then the output of the 4-bit.

DAC is subtracted from the V_{in} input and the remainder is fed into a second 4-bit
ADC converter that generates the lower 4 LSB of the conversion result. This architecture
requires two $2^4 - 1 = 2 \times 15 = 30$ comparators which is much less than the $2^8 - 1 = 255$
comparators required for a full 8-bit flash ADC architecture.

Fig. 6.15 An 8-bit half-flash
ADC block diagram

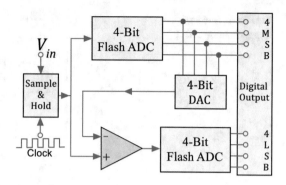

6.4.5 Successive Approximation ADC

In the successive approximation (SA) ADC a DAC cycles iteratively through a sequence of digital numbers while comparing the voltage level of each one of these numbers to the sampled signals at the input of the ADC. The sampled signal is kept constant during this process using a sample and hold circuit until conversion is completed. This occurs when the comparator op-amp indicates that one of the voltage outputs of the internal digital to analog converter (DAC) is equal (within the resolution) to the sampled signal V_{in}. Then the digital number input to the DAC at this stage will be accepted as the binary presentation for the sampled analog signal (V_{in}) at the input of the ADC. Digital inputs the internal DAC are generated by an n-bit register in a certain order controlled by the Control logic based on the output of the comparator (Fig. 6.16). The conversion sequence starts by setting the MSB (most significant bit) to 1 and all other bits to 0. If the DAC output corresponding to this binary code $V_{DAC} > V_{in}$, the comparator output will go low. The control logic will cause the MSB to be reset to 0. And the second bit to the right of MSB will be set to 1. Else if $V_{DAC} < V_{in}$, The MSB is kept at 1. In this second iteration, the next highest bit to the right is set to 1. Again, V_{DAC} is compared with V_{in}. If $V_{DAC} > V_{in}$, the second highest bit is reset to 0, otherwise it is kept at 1. This sequence continuous until the LSB (least significant bit) is processed as the previous ones. At this final stage the binary form in of the n-bit register is considered as the digital quantification of V_{in}. The analog signal is sampled periodically at a rate defined by the ADC characteristics.

A 4-bit converted will take 4 iterations (4 clock cycles). Each iteration will take one cycle. In general, it can be said, that for a n bit SA ADC it will take n × period of one clock cycle of conversion time which is independent of the magnitude of the input voltage. Conversion time will increase with increasing number of bits for better resolution and less quantization error. SA ADC is simple, it has a relatively low power consumption, and can achieve a conversion speed of up to 10 M samples/s.

Fig. 6.16 Successive
approximation ADC

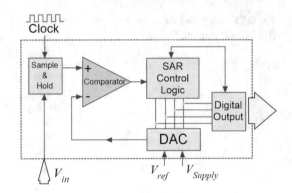

6.5   Examples

Example 6.1 Show the output of a 3-bit flash ADC with $V_{ref} = 8.0$ V for each of the
following three input signals $V_{in} = 6.3$ V, 2.5 V, and 0.5 V.

Solution
Referring to Fig. 6.14, the output of the priority encoder will be a binary representation
of the input with highest priority among the inputs with comparator's output High or 1.
The solution is presented in the following table.

Comparator No. (Fig. 6.14)	V_{in} 6.3 V	V_{in} 2.5 V	V_{in} 0.5 V
7	0	0	0
6	1	0	0
5	1	0	0
4	1	0	0
3	1	0	0
2	1	1	0
1	1	1	0
Encoder Output			
D2 (MSB)	1	0	0
D1	1	1	0
D0 (LSB)	0	0	0

Example 6.2 Determine the magnitude of the quantization error and its percentage relative to the full scale range for the two analog inputs shown in Fig. 6.2. When using a 4-bit ADC instead of a 3-bit ADC.

Solution

For the 0–10 V range, the quantization error is

$$Q_{Er} = \frac{V_{FSR}}{2^n} = \frac{10 \text{ V}}{2^4} = 0.625 \text{ V}$$

and is $\frac{0.625}{10} \times 100 = 6.25\%$ (FSR)

For the 0–5 V range, the quantization error is

$$Q_{Er} = \frac{V_{FSR}}{2^n} = \frac{5 \text{ V}}{2^4} = 0.3125 \text{ V}$$

and is $\frac{0.3125}{5} \times 100 = 6.25\%$ (FSR)

Example 6.3 Show the conversion sequence of a 4-bit successive approximation ADC for an input voltage of $V_{in} = 7.3$ V. This converter uses a reference voltage $V_{ref} = 10$ V.

Solution

As discussed in Sect. 6.4.5, the successive approximation ADC uses several iterations of trial-and-error to converge at the closest binary representation for the analog input voltage. The resolution is

$$Resolution = \frac{V_{FSR}}{2^n} = \frac{10 \text{ V}}{2^4} = 0.625 \text{ V}$$

With $V_{in} = 7.3$ V, The successive approximations are

1000 → 8 × 0.625 = 5 V < V_{in} (1st iteration)
1100 → 12 × 0.625 = 7.5 V > V_{in} → use 1010 (2nd iteration)
1010 → 10 × 0.625 = 6.25 V < V_{in} (3rd iteration)
1011 → 11 × 0.625 = 6.875 V < V_{in} (4th iteration) → Stop

The final 4-digit binary representation for 7.3 V is 1011 (Fig. 6.17) and is equivalent to a decimal voltage value of

$$\frac{V_{ref}}{2^4}(D_0 \times 1 + D_1 \times 2 + D_2 \times 4 + D_3 \times 8)$$

$$\Rightarrow V_{ref}\left(\frac{D_0}{16} + \frac{D_1}{8} + \frac{D_2}{4} + \frac{D_3}{2}\right)$$

$$\Rightarrow 10 \text{ V} \left(\frac{1}{16} + \frac{1}{8} + \frac{0}{4} + \frac{1}{2}\right) = 6.875 \text{ V}$$

which differs from $V_{in} = 7.3$ V. This is an example of the quantization error.

Example 6.4 Determine the digital display and the conversion time for a 4-digit multimeter (DMM) that uses an 8-bit single slope ADC operating on a 100 kHz clock frequency when connected to a 5.75 V signal source. Assume a converter reference voltage $V_{ref} = 10$ V.

Solution

Each counting step is equivalent to 1 bit (or LSB) which is equivalent to the resolution.

One integration step = Resolution = $\frac{V_{FSR}}{2^n} = \frac{10\,V}{2^8} = 0.03906$ V/step.

Or, 39.06 mV/step.

For a $V_{in} = 5.75$ V the required number of integration (conversion) steps are

Fig. 6.17 Decision tree for a 4-bit SA ADC (Example 6.3)

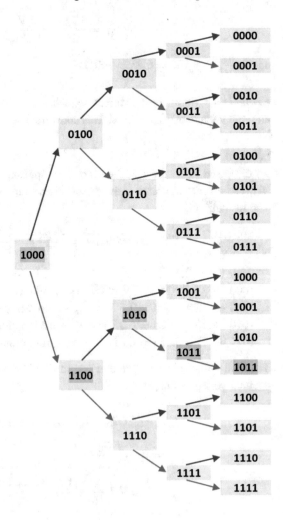

$$\frac{5.75\,\mathrm{V}}{0.03906\,\mathrm{V}/step} = 147.21 \; steps \approx 147 \; steps$$

The DMM display could be presented as

147 steps × resolution = 147 × 0.03906 = 5.742 V.

Each step takes one clock pulse period of (1/100,000) = 10 μs. Therefore, the conversion time is 147 × 10 μs = 0.00147 s = 1.47 ms.

Example 6.5 Assuming that, at start of conversion, all components' outputs in a sigma-delta ADC (Fig. 6.9) are set at 0, show the results for first 10 clock cycles taking for example V_{in} = 3.5 V, H = 5 V and L = 0 V. Assume unity gain for both the delta and sigma op-amps.

Solution

Following the discussion presented in Sect. 6.4.3. Table 6.1 illustrates the state at each component for each of the first 10 clock cycles.

Example 6.6 Determine the number of bits required for an ADC to achieve a resolution of 0.1% of the FSR, when operating within a FSR of 10 V.

Solution

Table 6.1 Sample signals data for a sigma-delta ADC

Clock	V_d	V_I	D	1-Bit DAC output	Average of D
0	0	0	0	0	
1	3.5	3.5	1	5	
2	−1.5	2	1	5	
3	−1.5	0.5	1	5	
4	−1.5	−1	0	0	
5	3.5	2.5	1	5	4
6	−1.5	1	1	5	
7	−1.5	−0.5	0	0	
8	3.5	3	1	5	
9	−1.5	1.5	1	5	
10	−1.5	0	0	0	3.5

$[V_{in} = 3.5\,\mathrm{V}, H = 5\,\mathrm{V}, L = 0\,\mathrm{V}]$

$$Resolution = \frac{V_{FSR}}{2^n} \Rightarrow \frac{Resolution}{V_{FSR}} = \frac{1}{2^n} = 0.001$$

$$\Rightarrow \ln(2^n) = \ln(1/0.001) \Rightarrow n = \frac{\ln(1/0.001)}{\ln 2}$$

$$\Rightarrow n = 9.97 \quad or, \quad n = 10$$

When using a 10-bit ADC the resolution would be

$$\frac{Resolution}{V_{FSR}} = \frac{1}{2^{10}} = 0.000977 \approx 0.098\%$$

Example 6.7 Determine the force resolution that would be achieved when using a force sensor with sensitivity of 4.0 mV/N and a full-scale output voltage of 0–5 V when employed with a 12-bit ADC that accepts the same voltage range as the sensor output.

Solution
The ADC resolution $= \frac{V_{FSR}}{2^n} = \frac{5\,V}{2^{12}} = 0.0012\ V = 1.2\ mV$
  Force resolution using ADC and force sensor is

$$\Delta F = \frac{1.2\ mV}{4.0\ mV/N} = 0.30\ N$$

6.6   Problems

6.1. A temperature sensor with and average sensitivity of 0.045 mV/°C at 0 °C is to be used w a 16-bit ADC. Determine the full scale voltage output that can be obtained with this setup.

6.2. Figure 6.18 shows a schematic for an 8-bit digital-to-analog converter (DAC) circuit. Answer the following questions regarding this converter.
   (a) Which switch connects the largest voltage to the op-amp? Why?
   (b) For 5 V reference voltage, determine the resistors values for R and R_f to produce a desired maximum output voltage range of 0–10 V.
   (c) Calculate the output voltage for an input digital word of 10,100,011.

6.3. Resolution is defined as the smallest change in the input signal that can be detected by the measurement system (sensor + DAQ). Determine the smallest change in input signal voltage that can be detected using a 12-bit ADC for a sensor with output voltage FSR (a) ± 5 V and (b) ± 0.2 V.

6.4. A dual slope ADC with R $= 100$ kΩ and C $=$ uses a 10 V reference voltage. If it uses a constant ramp time T_o of 5 ms, determine the total conversion time for an input voltage of 7.3 V.

Fig. 6.18 A schematic for an
8-bit DAC circuit

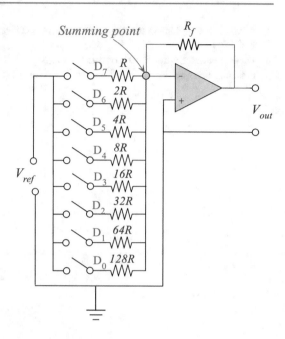

6.5. Given a 4-bit successive approximation ADC with a reference voltage of 10 V, determine the output of this converter for an input voltage of 3.450 V. (Show all conversion steps).

6.6. Find the decimal equivalent of the 8-bit binary 10,011,011.

6.7. Determine the digital display and the conversion time for a 4-digit measuring device that uses a 6-bit single slope ADC operating on a 30 kHz clock frequency when connected to a 3.24 V input signal. Assume a converter reference voltage $V_{ref} = 5$ V.

6.8. Discuss the operation sequence and the final binary value in the register and its equivalent output voltage for a 8-bit successive approximation ADC operating with a range of 0–5 V when the input is 3.250 V.

References

1. Issam A. Abu-Mahfouz, 'Instrumentation: Theory and Practice-Part II Sensors and Transducers.' Morgan and ClayPool Publishers, www.morganclaypool.com, 2022.
2. Horowitz, Paul, The art of electronics, 2nd ed., Cambridge University Press, New York, 1989.
3. Maurizio Di Paolo Emilio, Data Acquisition Systems: From Fundamentals to Applied Design, Springer, New York, NY, 2013

4. Alfi Moscovici, Understanding Data Converters Through SPICE. 1st ed., Springer, New York, NY, 2001.
5. Dieter Seitzer, Günter Pretzl, Nadder A. Hamdy, Electronic analog-to-digital converters: principles, circuits, devices, testing, J. Wiley, New York, 1983.

Printed in the United States
by Baker & Taylor Publisher Services